Colour of Inland and Coastal Waters

A Methodology for Its Interpretation

Springer
London
Berlin
Heidelberg
New York
Barcelona
Hong Kong
Milan
Paris
Santa Clara
Singapore
Tokyo

Dmitry Pozdnyakov and Hartmut Grassl

Colour of Inland and Coastal Waters

A Methodology for Its Interpretation

Springer

Published in association with
Praxis Publishing
Chichester, UK

PRAXIS

Professor Dmitry Pozdnyakov
Research Director
Nansen International Environmental and
Remote Sensing Centre
St Petersburg
Russia

Professor Hartmut Grassl
Director
The Max-Planck Institute
Hamburg
Germany

SPRINGER–PRAXIS BOOKS IN MARINE SCIENCE AND COASTAL MANAGEMENT
SUBJECT *ADVISORY EDITOR*: Selina Stead, B.Sc., M.Sc., Ph.D., Programme Director for Marine Resource
Management, University of Aberdeen, UK

ISBN 3-540-00200-6 Springer-Verlag Berlin Heidelberg New York

Die Deutsche Bibliothek – CIP-Einheitsaufnahme
A catalogue record for this book is available from the Deutsche Bibliothek

Library of Congress Cataloging-in-Publication Data
A catalogue record for this book is available from the Library of Congress

Cover design: Jim Wilkie
Project Management: Originator Publishing Services, Gt Yarmouth, Norfolk, UK

Printed on acid-free paper supplied by Precision Publishing Papers Ltd, UK

Contents

Preface

The propagation of downwelling direct solar and diffuse radiation through natural water bodies results not only in elastic scattering and absorption by water molecules but also in a variety of photon interactions with the co-existing water constituents of different nature. The radiative transfer mechanisms governing the spectral composition of the light emerging from the water surface ultimately determine the water colour as it is perceived by a human eye. The inorganic and organic water constituents, often called *colour producing agents* (CPAs), responsible for water colour are generally referred to as *water quality parameters*. Therefore, the water colour can be considered as a convolution of light interactions with CPAs and H_2O molecules, and as such is not infrequently exploited as a quality characterizing the current ecological/sanitary/resources status of the aquatic environment.

Utilization of water colour for assessment of water quality parameters can be achieved through the application of the accomplishments in aquatic optics attained over many decades. Indeed, an adequate knowledge of photon propagation in natural water, as a scattering and absorbing medium, can serve as a basis for development of procedures (algorithms) that would give a scientifically justifiable meaning to the optical data collected from above the water–air interface, and a means of retrieving the concentrations of CPAs.

Aquatic optics can be subdivided according to whether the natural water body is salty (oceanic), inland or fresh (limnological), or coastal (often brackish). The degree of optical complexity of a natural water body/mass, and hence the description/ modelling of its interactions with visible light, is, in general, related to its proximity to land masses.

A. Morel and L. Prieur (1977) have suggested a bipartite division of the world oceans waters, according to which all natural waters are either Case I or Case II waters. In Case I waters, phytoplankton together with accompanying and co-varying products of their life cycles, as well as some microscopic organisms such as flagellates, bacteria and viruses (which are also indigenous to offshore/mid-oceanic waters), are the principal agents determining the variations in optical properties of sea water. If present, the substances other than phytoplankton are generally relatively scarce, and the optical properties of Case I waters can thus be modelled in most cases just as a function of phytoplankton concentration (Sathyendranath, 2000).

The optical properties of Case II waters, unlike Case I waters, are influenced not only by phytoplankton and the substances originating from the phytoplankton's life cycle evolution, but also by other substances generated independently of phytoplankton, notably inorganic/terrigenous particulate matter in suspension and dissolved organics. Their content in the water column is often abundant enough to compete with phytoplankton in influencing the resultant optical properties of Case II waters, thus rendering such waters optically very complex. This optical complexity of Case II waters rapidly escalates when approaching the coast. Naturally, it progressively complicates the task of interpreting water colour with the goal of inferring the composition and concentrations of CPAs. In other words, the task of remote sensing becomes, under such conditions, extremely complicated if not frustrating.

It is further aggravated by the fact that unlike oceanographic optics, which has long since benefited from rapid advances in the development of sophisticated submersible spectrometric instrumentation, limnological/coastal optics has not enjoyed the same quality and variety of hydro-optical devices with the appropriate spectral resolution within the appropriate spectral region, the dynamic range of sensitivity, and the required vertical resolution of measured quantities. This lack of dedicated instrumentation has hindered and still is hindering progress in acquiring/collecting *in situ* data on the optical properties of Case II waters.

Moreover, the optical complexity of Case II waters (both in terms of their composition and characteristic abundance of CPAs) implies in many cases the necessity of considering a number of mechanisms of photon interactions with molecules and particles residing in the water column, as well as with the bottom (in the case of optically shallow waters), which are less important when dealing with Case I waters. It requires, along with multifaceted *in situ* hydro-optical data, a development of new theoretical approaches to account for the impacts of such interactions (e.g. fluorescence) on the upwelling radiance and, hence, the water colour perceived by a remote observer.

This, in turn, entails the necessity of applying/developing new, much more sophisticated techniques for the retrieval of CPAs. Furthermore, when Case II waters are heavily laden with CPAs, the upwelling radiance can be very weak and account for only 1–2% (sometimes even less) of the radiation captured by a satellite sensor. The rest of the captured signal is mostly due to the contribution of the intervening atmosphere (often called *path radiance*) and, to a lesser degree, the diffuse sky light reflected from the water surface, sun glint being automatically avoided by appropriate tilting of the sensor. It implies extremely high accuracy measurements for a proper removal of both path radiance and surface reflection.

Finally, typical time and space scales of features to be monitored in coastal and inland water bodies are often so specific that a simple adaptation of sensors built for Case I water environments will not meet the challenges associated with the remote sensing of Case II water resources. A successful remote-sensing mission in Case II waters necessitates the use of specialized satellite sensors designed to meet both the ground resolution and repeat cycle requirements arising from the nature of the phenomena to be studied/monitored, the spatial and temporal resolution of complementary *in situ* measurements (that remain necessary and should be carefully designed) as well as the end-users' needs.

In view of all these methodological 'perils', shortages in *in situ* instruments, and the need for specialized satellites to successfully conduct a remote-sensing mission in Case II waters, a logical question arises: is it worth all the trouble?

Even though Case II waters constitute only a small fraction (about 8%) of the total water-covered area of our planet, their economic, social and ecological significance cannot be overlooked. Sixty per cent of the human population lives in the coastal zone, defined by Pernetta and Milliman (1995) as 'the region from 200 m above to 200 m below mean sea level'. Two-thirds of the world's large cities are coastal ones. Coastal waters account for 14% of global ocean primary production, 90% of world fish catch, 75–90% of global sink of suspended river load (Parslow *et al.*, 2002). Human recreational activities and tourism, important habitats, coral reefs' inclusive and rapidly growing mariculture, far from being a complete list of the significance of coastal and inland waters, warrant the efforts at national and international levels to achieve sustainable management of Case II water resources. Remote sensing can and must play an important role here in timely surveillance and diagnosis of current trends/dynamics in the ecological state of coastal and inland water bodies.

There are yet other reasons warranting a thorough research in the entire realm of Case II waters problems incorporating hydro-optics, hydrobiology and extraction of the useful signal at satellite level. In fact, it is recognized by many workers (for references see Sathyendranath, 2000) that phytoplankton are not necessarily the only agents determining the spectral distribution of upwelling radiance, and hence the colour in Case I waters. For instance, the results obtained in oligotrophic tropical areas suggest that dissolved organics may be more abundant, and more variable than hitherto believed (Church *et al.*, 2002) and their optical impact is not negligible, if not comparable with that caused by phytoplankton (Bouman *et al.*, 2000). It naturally raises the question whether such offshore/pelagic oceanic waters, which traditionally have been considered as Case I waters, indeed always belong to this category. Can sporadic and/or periodic switching to Case II waters occur in mid-oceanic (Case I) areas? If this holds then the CAP retrieval algorithms, or at least the conceptual approaches developed for Case II waters, may eventually be applicable to Case I waters as well. Indeed, in this case the strict prerequisite of co-variance between phytoplankton and the accompanying suite of phytoplankton life-cycle-related biological substances can be relaxed even in Case I waters with a possible increase in the retrieval accuracy of phytoplankton concentration. Then, perhaps, it is not unreasonable to foresee the development of water quality retrieval algorithms that are applicable to both Case I and Case II waters.

In the present book we have aimed to reach two major goals. The first one relates to the solution of the forward problem in the case of both coastal or inland waters, and pure pelagic waters. That is, we investigate the formation/spectral composition of radiance upwelling from beneath the water surface. In doing so, we take into account not only absorption and elastic scattering of solar radiation in the water column, but also, when appropriate (i.e. in optically shallow waters) its reflection at the bottom. Simultaneously, the so-called trans-spectral processes (such as Raman scattering by water molecules and fluorescence by phytoplankton chlorophyll and dissolved organics) are equally taken into account. The optical impact of the totality of these interactions is explored in terms of water colour variations as a function of varying CPA concentrations, sun zenith angle, and bottom cover.

The second goal consists in attaining a solution of the inverse problem, which in the case of remote sensing is reduced to the retrieval of CPA concentrations from the upwelling radiance, or water colour. A number of retrieval algorithms are exploited to this end and analysed in terms of their robustness under conditions of significant spatial/temporal heterogeneity of CPAs in natural waters. The problem is first treated in the absence of the atmosphere and under calm water surface conditions. Then, the effects of water roughness and atmospheric interference corrupting the useful signal are discussed.

The final chapter of the book is dedicated to our experience in practical application of the developed methodologies to processing SeaWiFS images of some large water bodies with a pronounced heterogeneity of their optical properties.

This book is conceived by the authors as a way of sharing mostly our own experience in coping with the complex problem of sounding non-Case I waters and interpreting the collected optical data, rather than as a textbook exhaustively providing the theoretical fundamentals of both aquatic optics and remote sensing. We apologize to those readers who expected to find in this book a consistent and profound discussion of light transfer theory and the relevant mathematical formulations. Only a concise synopsis of both basic notions in hydro-optics/radiometry and the formulism used by us in solving the forward problem can be found herein. We have not tried either to give a voluminous review of what has been done so far in the area of Case II waters: only those papers are discussed/cited here, on which we either based some of our simulations or found as substantiating our results and/or corollaries. To those who expected to find in this book quick and easily attainable solutions to the problems of remote sensing of optically complex waters we apologize for failing to suggest simplistic recipes/approaches. To those who accept the nature of our endeavour, we bid you welcome to *Colour of Inland and Coastal Waters: A Methodology for Its Interpretation.*

D. Pozdnyakov

H. Grassl

Acknowledgements

The authors acknowledge with thanks the financial support provided by the European Commission under the projects INCO-COPERNICUS (ICA2-CT-2000-10014 'WHITESEA') and INTAS (INTAS-99-674 'FINGULF'): part of the research conducted under these extensive projects constitutes the content of this book.

The authors also express their profound gratitude to Drs Anton Lyaskovsky and Robert Bukata: parts of this monograph are based on the research conducted jointly with them.

About the authors

Dmitry Pozdnyakov, Ph.D., Prof. Dr, received his doctorate in physics in 1972 from the State University of St. Petersburg, Russia, where he conducted infrared studies simulating gas/aerosol interactions. His work revealed new sink mechanisms of climate-controlling gases, including the stratospheric ozone-depleting fluorocarbons. He then accepted a lecturing post in the Department of Atmospheric Physics at the State University. As a Visiting Professor he lectured in physics and atmospheric optics for five years at the University of Conakry, Guinea.

In 1983 he joined the Institute for Lakes Research of the Russian (then USSR) Academy of Sciences, where his research interests focused upon limnological ecology and hydro-optics. He has developed bio-optical algorithms for remote sensing of water quality parameters utilizing passive spectrometric and active lidar techniques. His scientific team has remotely investigated the trophic status of nearly all the large lakes and water storage reservoirs of the former USSR.

He was awarded a D.Sc. degree in 1992, and in 1996 he became a full University professor. He authored more than one hundred scientific papers, brochures, and books. He currently holds the position of Research Director at the Nansen International Environmental and Remote Sensing Centre (NIERSC), and is Invited Professor at the Electrotechnical University, both in St. Petersburg. His research activities continue to be directed towards the optical properties of inland, coastal and marine waters.

In 1992 he and R. P. Bukata, J. H. Jerome and K. Ya. Kondratyev were recipients of the Chandler–Misener Award presented by the International Association for Great Lakes Research for their work documented in a number of companion publications.

Hartmut Grassl, Ph.D., Prof. Dr, received his doctorate in Meteorology in 1970 from the University of Munich, Germany, where he worked on atmospheric spectroscopy in the near- and mid-infrared spectral range. During a postdoctoral phase in Mainz, Germany, he was one of the first to explore with the help of new numerical codes how the radiative transfer through clouds is influenced by aerosols. In recognition of this pioneering work he was granted the Habilitation in 1978 by the Universität Hamburg, Germany, where he continued his contributions to the understanding of the influence of cloud on climate. During his time as Professor for Theoretical Meteorology at the Institut für Meereskunde, Kiel, Germany, between 1981 and 1984, he started intensive work on the remote sensing of atmospheric and ocean surface properties through the analysis of thermal infrared satellite data. He continued this topic during his engagement as the Director of the Institute of Physics at the GKKS Research Centre in Geesthacht near Hamburg, where he also supervised working groups on regional atmospheric modelling, on the modelling and prediction of ocean wave spectra and on remote sensing of sea water constituents.

In 1988 he was appointed a member of the German Max-Planck-Society, became full Professor for Meteorology at the University of Hamburg, and at the same time one of the directors of the Max-Planck-Institut für Meteorologie in Hamburg. In this position he directed several scientific teams to tackle and improve work on important climate-relevant processes and remote sensing methods, including radar and lidar techniques for the active probing of the lower troposphere. From 1988 to 1994 he contributed to the Intergovernmental Panel on Climate Change (IPCC) and from 1989 to 1994 he was a member of two Enquiry Commissions of the German Parliament on the 'Protection of the Earth's Atmosphere'. Between 1992 and 1994 he acted also as the chair of the Global Change Advisory Council to the German Government. In 1994 he was selected as the director of the World Climate Research Programme (WCRP) in Geneva, Switzerland. In this position he served as a coordinator of international global change research and engaged himself strongly in spreading knowledge of the results in order to persuade mankind and their leaders to take appropriate mitigation measures soon. Since 1999 he has been back in his faculty position and as a director of the Max-Planck Institute in Hamburg, continuing his research activities and his endeavours for further development of global change science and public awareness of related problems.

He has authored more than 200 scientific papers, review articles and books. He is currently managing editor of the journal *Theoretical and Applied Climatology*. He has received several prestigious awards, among them the Young Scientist Award of the German Meteorological Society in 1971, the Max-Planck-Prize in 1991, Das große Bundesverdienstkreuz (German Order of Merit) in 2002 and the German Environmental Prize in 1998.

Figures

Tables

Colour plates (between pages 138–139)

1

Basic notions and relationships

1.1 WATER COLOUR

From the perspective of the human observer, colour is the result of the interplay between the light spectrum reaching the eye and the spectral response of the eye's retina. According to the Young–Helmholtz theory, the human colour perception is normally trichromatic, i.e. responsive to the three spectral regions red, green, and blue. Therefore, any perceived colour is a result of appropriate partitioning between red, green, and blue light. This thesis constitutes the basis of chromaticity analyses.

Integration of the sensitivity of the human eye with the spectrum of the radiant flux per unit area of an extended surface impinging upon it (otherwise called *irradiance*, E, see Section 1.2) results in *tristimulus* values X', Y', and Z', from which the chromaticity coordinates X (red), Y (green), and Z (blue) may be determined. A measured upwelling irradiance $E(\lambda)$ may therefore be related to a perception of visual colour of the water column.

The tristimulus values of an upwelling irradiance spectrum $E(\lambda)$ are defined (Anonymous, 1957) as:

$$X' = \int E(\lambda)x(\lambda)\,\mathrm{d}\lambda, \qquad (1.1)$$

$$Y' = \int E(\lambda)y(\lambda)\,\mathrm{d}\lambda, \qquad (1.2)$$

$$Z' = \int E(\lambda)z(\lambda)\,\mathrm{d}\lambda, \qquad (1.3)$$

where $x(\lambda)$, $y(\lambda)$, and $z(\lambda)$ are CIE *colour mixture* data for the red, green, and blue regions of the spectrum, respectively. The numerical values of $x(\lambda)$, $y(\lambda)$, and $z(\lambda)$ are those corresponding to an equal energy spectrum, and are listed, for example, in Table 6.1 in Bukata *et al.* (1995).

The chromaticity coordinates X, Y, and Z then can be obtained from:

$$X = X'/(X' + Y' + Z'), \tag{1.4}$$

$$Y = Y'/(X' + Y' + Z'), \tag{1.5}$$

$$Z = Z'/(X' + Y' + Z'). \tag{1.6}$$

Chromaticity coordinates may be computed for each λ through the entire visible spectrum. Since, understandably, the sum of the three chromaticity coordinates is unity, only two chromaticity coordinates are required to plot a chromaticity diagram relating corresponding pairs to one another. The locus of the plotted pairs defines an envelope, which encompasses all possible chromaticity values, and hence colours, perceivable by the human eye. For polychromatic or 'white' light, $X = Y = Z = 0.333$. The location of this point within a chromaticity diagram defines the white point or the achromatic colour S (Fig. 1.1).

A numerical value ascribed to colour is then obtained geometrically by drawing a line from this white point S through the plotted pair of chromaticity coefficients of a measured upwelling spectrum (indicated in the figure by the point Q) and intersecting the chromaticity envelope. This intersection of the line SQ with the chromaticity envelope (indicated by the point A) specifies the dominant wavelength (λ_{dom}) of the observed irradiance spectrum, considered as a colorimetric definition of the

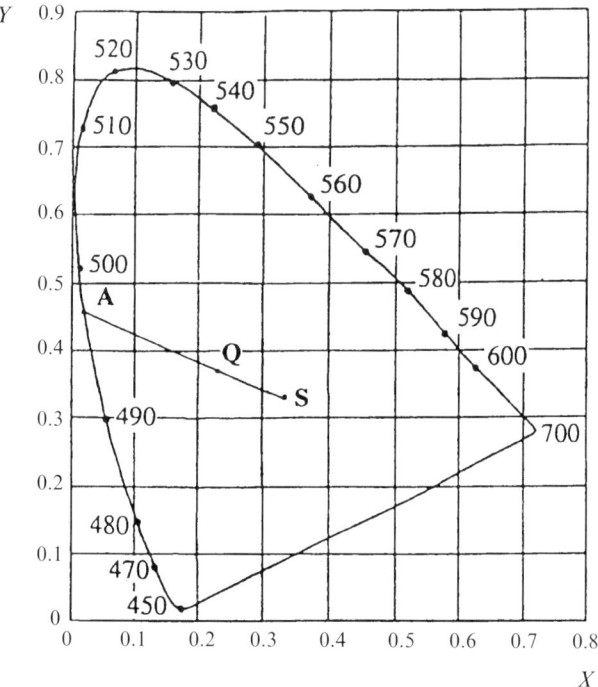

Figure 1.1. Chromaticity diagram using the chromaticity coordinates Y (green) and X (red).

perceived colour. The dominant wavelength defines the monochromatic light, which has the colour with the chromaticity coordinates corresponding to this specific point on the chromaticity diagram.

Such a dominant wavelength definition of perceived colour applied to natural water bodies, however, encounters difficulties: under real conditions the perceived water colour proves to be 'contaminated' with the sunlight elastically scattered back into the atmosphere by the aquatic medium. This results in displacing the point Q along the line AS. The distinctiveness of the observed water colour may be attained through involving an additional characteristic termed spectral purity (p), which is a measure of the magnitude of the actual contribution of the monochromatic light to the perceived colour. A spectral purity of 1.0 indicates that the observed colour is completely composed of monochromatic light with the dominant wavelength corresponding to the point A. A spectral purity p of 0 indicates a 'white' spectrum. Thus, the dominant wavelength λ_{dom} and its associated spectral purity p uniquely define an aquatic colour.

1.2 CHARACTERIZATION OF THE LIGHT FIELD IN WATER

As was emphasized in the Preface, the major thrust of this book will be directed towards an interpretation of visible light (that is, radiation contained within the boundaries of the colour wavelengths) emanating from inland and coastal natural water bodies. To accommodate this thrust, the basic notions and relationships relevant to aquatic optics and light propagation will now be briefly reviewed.

Sun photons entering and propagating within a natural water body will undergo scattering and absorption interactions with those constituents of the natural water body that can qualify as *colour producing agents* (CPAs). Both absorption and scattering reduce the intensity of the radiance distribution, while the scattering processes also change the directional/angular distribution of the propagating radiation.

Radiance and *irradiance* are the two most important terms required for understanding the behaviour of a photon flux. The radiance, L, in a specified direction at a point in the radiation field is defined as the radiant flux Φ at that point per unit solid angle $d\Omega$ per unit area dA at *right angles* to the photon propagation. The standard unit of radiance is the watt per square metre per steradian ($\mathrm{W\,m^{-2}\,sr^{-1}}$). The irradiance E refers to the radiant flux Φ per unit area at a point within the radiative field integrated over all directions (solid angle Ω) of a hemisphere. Remembering that in polar coordinates $d\Omega = \sin\theta\,d\theta\,d\phi$, we can write:

$$E = \int_{\Omega} L(\theta, \phi) \cos\phi\,d\Omega \equiv \int_{0}^{2\pi}\int_{0}^{\pi/2} L(\theta, \phi) \cos\theta \sin\theta\,d\theta\,d\phi. \qquad (1.7)$$

Accordingly, the unit of irradiance is $\mathrm{W\,m^{-2}}$.

Hence, depending on the direction of the flux, irradiance can be treated as a directional parameter with *upwelling* and *downwelling* irradiances being considered as distinguishable entities. *Downwelling* irradiance can be obtained by integrating the radiance over the upper hemisphere:

$$E_d = \int_0^{2\pi} \int_0^{\pi/2} L(\theta, \phi) \cos\theta \sin\theta \, d\theta \, d\phi, \qquad (1.8)$$

where θ and ϕ are zenith and azimuth angles, respectively.

Analogously, *upwelling* irradiance is the result of the radiance integration over the lower hemisphere (with due consideration that $\cos\theta$ is negative for $\pi/2 \le \theta \le \pi$):

$$E_u = -\int_0^{2\pi} \int_{\pi/2}^{\pi} L(\theta, \phi) \cos\theta \sin\theta \, d\theta \, d\phi. \qquad (1.9)$$

At any point within the water column, the *net* downwelling irradiance $E \downarrow$ is the algebraic sum of E_d and E_u : $E \downarrow = E_d - E_u$.

It is often appropriate to consider the total radiant intensity at a point in space independent of its arrival directions. It implies regarding radiance from each direction equally. This, in turn, is equivalent to the removal of the $\cos\theta$ term from eqs (1.7) and (1.8). The integration of such equally treated radiances results in the irradiance at that point termed the *scalar irradiance*, E_0:

$$E_0 \int_0^{2\pi} \int_0^{\pi} L(\theta, \phi) \sin\theta \, d\theta \, d\phi. \qquad (1.10)$$

The unit of E_0 is $\mathrm{W\,m^{-2}}$. Analogously with partitioning $E \downarrow$ into E_d and E_u, the net scalar irradiance can also be represented as an algebraic sum of downwelling and upwelling scalar irradiances: $E_0 \downarrow = E_{0d} - E_{0u}$:

$$E_0 \downarrow = \int_0^{2\pi} \int_0^{\pi/2} L(\theta, \phi) \sin\theta \, d\theta \, d\phi - \int_0^{2\pi} \int_{\pi/2}^{\pi} L(\theta, \phi) \sin\theta \, d\theta \, d\phi. \qquad (1.11)$$

Two properties of the angular distribution of light are commonly used: the mean cosine for downwelling irradiance, $\mu_d = E_d / E_{0d}$, and the mean cosine for the upwelling irradiance, $\mu_u = E_u / E_{0u}$. The average value of the mean cosine, μ, of zenith angle θ for the total 4π radiance distribution at a point, is the *net* downwelling irradiance normalized to the *net* downwelling scalar irradiance:

$$\mu = E \downarrow / E_0 \downarrow = \frac{E_d - E_u}{E_{0d} - E_{0u}}. \qquad (1.12)$$

1.3   INHERENT AND APPARENT OPTICAL PROPERTIES

Numerically, the attenuation of direct sunlight (as opposed to diffuse sky light) in the water body can be assessed via the beam absorption coefficient $a(\lambda)$, beam scattering coefficient $b(\lambda)$, and beam attenuation coefficient $c(\lambda)$, with:

$$c(\lambda) \equiv a(\lambda) + b(\lambda), \qquad (1.13)$$

where λ, as above, is the wavelength of the light propagating through the aquatic medium, and a, b, c are given in m^{-1}.

Importantly, the values of $a(\lambda)$, $b(\lambda)$, and $c(\lambda)$ are independent of the manner in which the medium under observation is being illuminated. Such optical properties, therefore, should be independent of the radiation distribution within this medium and entirely depend upon the intrinsic physical/chemical attributes of the water constituents. Accordingly, all three quantities $a(\lambda)$, $b(\lambda)$, and $c(\lambda)$ qualify as *inherent optical properties* (IOPs) of the medium.

It is noteworthy that the inherent optical properties are *additive* in nature and result from the individual contributions of the co-existing absorbing and/or scattering constituents indigenous to natural waters:

$$a = \sum_{i}^{N} a_i; b = \sum_{j}^{N'} b_j; b_b = \sum_{j}^{N'} b_{b_j},$$

where $i = 1, 2, 3, \ldots$ and N; $j = 1, 2, 3, \ldots, N'$; N and N' are the total numbers of absorbing and scattering constituents co-existing in water.

The above additive expressions for a, b, and c can be further modified via introducing *specific* absorption a^* and *specific* scattering b^* (backscattering b_b^*) coefficients (also referred to as absorption and scattering (backscattering) *cross-sections*):

$$a = \sum_{i}^{N} C_i a_i^*; b = \sum_{j}^{N'} C_j b_j^*; b_b = \sum_{j}^{N'} C_j b_{b_j}^*,$$

where i, j, N, and N' are as above, and C_i and C_j are the concentrations of absorbing and scattering agents, respectively.

The angular distribution of the scattered light resulting from the aforementioned photon interactions is specified in terms of a *volume scattering function*, $\beta(\theta, \phi)$, where θ is the zenith angle and ϕ the azimuthal angle. $\beta(\theta, \phi)$ defines the ratio of the scattered radiance in a direction (θ, ϕ) per unit scattering volume to the value of the incident irradiance E. Assuming the independence of azimuth (which is a common approximation in the majority of natural waters), the angular dependence of the volume scattering function is generally confined to θ. If normalized to the scattering coefficient b, the volume scattering function $\beta(\theta)$ is named the *scattering phase function*, $P(\theta)$. Understandably, in the case of an *isotropic* scattering medium, the phase function P is equal to unity. Together with a, b and c, $\beta(\theta, \phi)$ and $P(\theta)$ are also IOPs.

Since the light emerging from beneath the water surface (and eventually captured by a remote sensor) consists of photons reflected within the water column into the hemisphere facing (although, generally, in a skewed manner) the water–air interface, it is meaningful to single out the *backscattering* coefficient, $b_b = Bb$ along with the *forward* scattering coefficient $b_f = Fb$, where B and F are the *backscattering* and *forward scattering probability*, respectively. Since all the photons scattered by a unit volume are either scattered forward or backward, we get

$$b_f + b_b = b. \tag{1.14}$$

The direct (sun)light propagation through the water column is accompanied by downward sky (*diffuse*) light. The irradiance beneath the air–water interface is attenuated as the light propagates downwards through the water column. The downwelling irradiance attenuation is described by the *irradiance attenuation coefficient* $K_d(\lambda, z)$, the logarithmic depth derivative of the spectral irradiance at subsurface depth z:

$$K_d(\lambda, z) = -\frac{1}{E_d(\lambda, z)}\left[\frac{\partial E(\lambda, z)}{\partial z}\right], \tag{1.15}$$

given in m^{-1}.

This definition (1.15) is consistent with Beer's Law as integration over z downwards from the air–water interface yields:

$$E_d(\lambda, z) = E_d(\lambda, -0)\exp[-K_d(\lambda, z)z], \tag{1.16}$$

where $E_d(\lambda, z)$ and $E_d(\lambda, -0)$ are the values of the downwelling irradiance at depth z and just below the air–water interface, respectively.

The intrinsic colour of a water body is defined by spectral variations in *volume reflectance*, R, just beneath the water surface, the ratio of upwelling irradiance to downwelling irradiance at the same level within the water column. For $z = -0$:

$$R(\lambda, -0) = \frac{E_u(\lambda, -0)}{E_d(\lambda, -0)}. \tag{1.17}$$

Both volume reflectance, R, and irradiance attenuation coefficient, K, qualify as *apparent optical properties* (AOPs) since unlike IOPs, they are dependent upon the spatial distribution of incident radiation.

A number of parameterizations have been suggested hitherto to relate AOPs to IOPs (Gordon *et al.*, 1975; Gordon, 1989; DiToro, 1978; Kirk, 1981, 1984, 1991; Sathyendranath and Platt, 1997; for a detailed discussion see Bukata *et al.*, 1995, and Kondratyev *et al.*, 1999). Analytical investigations as well as numerical simulations and semi-empirical approaches have been used to this end.

For instance, through Monte Carlo simulations, Jerome *et al.* (1988a,b) obtained for a calm water surface the following relationships between R and OAPs:

$$R(\lambda, -0, \theta_0') = (1/\mu_0)\,0.319 b_b(\lambda)/a(\lambda) \qquad \text{for } 0 \leq b_b/a \leq 0.25, \tag{1.18}$$

and

$$R(\lambda, -0, \theta_0') = (1/\mu_0)[0.013 + 0.267 b_b(\lambda)/a(\lambda)] \qquad \text{for } 0.25 \leq b_b/a \leq 0.5, \tag{1.19}$$

where $\mu_0 = \cos(\theta_0')$, with θ_0' being the in-water sun zenith refracted angle. $\mu_0 = 0.858$ is taken for overcast conditions.

Numerous field experiments in the Great American Lakes as well as in other water bodies suggest that eqs (1.18) and (1.19) hold for sun zenith angle $\theta_0 \leq 55°–60°$ (Bukata *et al.*, 1995). This conclusion has been confirmed through dedicated studies conducted in the Russian large lakes such as, for example, Lakes Ladoga and Onega (Kondratyev *et al.*, 1999). Eqs (1.18) and (1.19) are robust also in waters with a pronounced spatial heterogeneity of optical properties. The dependence of R on

$1/\cos\theta_0$ (and hence on the volume scattering function, β, of the aquatic medium) was investigated in dedicated research projects, and the interested reader can find the appropriate reports, for example, in Gordon (1989) and Kirk (1991; 1999).

In the context of remote sensing, it is more appropriate to use the so-called remote-sensing reflectance, $R_{rs}(\lambda, +0)$, defined as the upwelling radiance leaving the water surface, $L_u(\lambda, +0)$, normalized by the downwelling irradiance, $E_d(\lambda, +0)$, just above the water surface. It has the dimension [sr^{-1}]. $E_d(\lambda, +0)$ is the sum of diffuse, $E_{d,sky}(\lambda, +0)$ and direct $E_{d,sun}(\lambda, \theta_0, +0)$ solar irradiance.

The remote sensing reflectance, $R_{rs}(\lambda, +0)$, which can be directly determined from atmospherically corrected satellite data, is closely related to volume reflectance of the water body:

$$R_{rs}(\lambda, +0) = Q^{-1} R(\lambda, -0), \qquad (1.20)$$

where Q (in sr) is the ratio of subsurface upwelling irradiance $E_u(\lambda, -0)$ to subsurface upwelling radiance $L_u(\lambda, \theta'_v, -0)$, with the viewing angle θ'_v (just beneath the water surface).

The major difficulty in relating $R_{rs}(\lambda, +0)$ and R, resides in the absence of a good estimate of Q for each concrete sounding (Morel et al., 1995; Morel and Gentili, 1996; Jerome et al., 1996; Zibordi and Berton, 2001; Loisel and Morel, 2001). Indeed, Q depends not only on the solar zenith angle, θ_0, but also on the backscattering probability B or scattering phase function $P(\theta)$ and wavelength λ: $Q < 3.0$ sr holds at low sun zenith angles, but it can be as high as nearly 6 sr at large solar zenith angles and low B (Jerome et al., 1988b).

Prior to Jerome et al. (1988b), Siegel (1984) found from simulations that the value of Q ranged from 3.4 to 6.4 sr for the viewing angle $\theta_v = 0$, and is almost independent of wavelength in the spectral interval 450–650 nm, but has a strong (exponential) dependence on θ_0 in the range from 25° to 80°. The latter was confirmed in field experiments conducted by Aas and Hojerslev (1999) in the Mediterranean Sea. Measurements conducted in Lake Pend Oreille have yielded Q values varying between about 1 and 5 sr (Morel et al., 1995).

Morel and Gentili (1996) have also found that for clear oceanic waters, i.e. Case I waters, Q increases with θ_0, but also it does with wavelength λ, and the concentration of chlorophyll (C_{chl}): at $\lambda = 670$ nm, nadir view, $\theta_0 = 75°$ and $C_{chl} = 3$ mg m^{-3} Q reaches 5.7–5.8 sr.

For Case II waters, Loisel and Morel (2001) have assessed mean values of Q as a function of θ_0 for two distinctly different cases, viz. sediment-dominated waters (S-2) and yellow substance-dominated waters (Y-2). The volume scattering function was derived from the Petzold measurements (Petzold, 1972). For the 400–600 nm spectral region the mean Q(S-2) and Q(Y-2) values at $\theta_0 = 0°$ are equal to 3.53 sr and 3.69 sr, respectively. They progressively grow with θ_0, and at $\theta_0 = 75°$ reach 4.09 and 5.02 sr, respectively.

Recently Mobley et al. (2001) analysed the phase function effects on light fields in the ocean and concluded that for Case II waters the use of a depth- and wavelength-dependent Fournier–Forand phase function (Fournier and Forand, 1994) gave much better agreement with measured downwelling and upwelling irradiances than did commonly used phase functions from Petzold (1972), whose backscatter fraction was too large. This finding adds even more uncertainty to the adequacy of the aforementioned modelled Q-values.

As seen, the factor Q, being a function of so many parameters, is highly variable in natural waters and is not known *a priori*. This creates difficulties in using $R_{rs}(\lambda, +0)$ for the solution of the inverse problem, i.e. the retrieval of concentrations of water constituents (CPAs) from optical remote sensing data corrected for atmospheric effects.

To avoid these difficulties associated with the actual value of Q, an attempt was made to relate $R_{rs}(\lambda, +0)$ just above the surface to R_{rsw} just beneath the air–water interface. When neglecting the contribution of internally reflected upwelling radiance to $E_d(\lambda, -0)$, R_{rs} and R_{rsw} are related by

$$R_{rs}(\theta_0) = R_{rsw}(\theta_0')[(1 - \rho_{surf})(1 - \rho_{int}(\theta_0'))/n^2], \qquad (1.21)$$

where ρ_{surf} is the surface reflectivity for incident downwelling irradiance (a function of both θ_0 and the fraction F of diffuse sky irradiance in the incident irradiance, see below), ρ_{int} is the internal surface reflectivity for an incident angle θ_0' (θ_0' is the in-water solar zenith angle, related to in-air solar zenith angle θ_0 through Snell's law), and n is the relative index of refraction (≈ 1.333). It was shown by Jerome *et al.* (1988b) that:

$$\rho_{int}(\theta_0') = 0.271 + 0.249\mu_0', \qquad (1.22)$$

where $\mu_0' = \cos(\theta_0')$ and ρ_{int} is equal to 0.561 for a cardioidal diffuse incident radiation distribution. Values of ρ_{surf} for direct *unpolarized* solar radiation can be calculated from Fresnel's equation (Jerlov, 1976):

$$\rho_{sun}(\lambda, \theta_0) = 1 \left/ 2 \left[\frac{\sin^2(\theta_0 - \theta_0')}{\sin^2(\theta_0 + \theta_0')} + \frac{\tan^2(\theta_0 - \theta_0')}{\tan^2(\theta_0 + \theta_0')} \right] \right., \qquad (1.23)$$

while ρ_{surf} for a diffuse *cardioidal* distribution can be taken as 0.066 (Jerlov, 1976). The values of ρ_{surf} (for direct and diffuse incident radiation) as well as of ρ_{int} are also discussed by Baker and Smith (1990).

Jerome *et al.* (1996) have found through Monte Carlo simulations that the *subsurface remote sensing reflectance*, R_{rsw}, can be related for nadir view, flat water surface, a very wide range of in-water optical conditions (the backscattering probability varied between 0.0133 and 0.0440), sun zenith angles θ_0 from $15°$ to $89°$, to the inherent optical properties (IOPs) at high correlation (0.99) and an rms error of only 9% by:

$$\overline{R_{rsw}} = -0.00036 + 0.110(b_b/a) - 0.0447(b_b/a)^2. \qquad (1.24)$$

For an isotropic in-water and a cardioidal incident radiance distribution:

$$R_{rsw} = 1.045\overline{R_{rsw}}. \qquad (1.25)$$

This approach circumvents the difficulties associated with Q, and can be used for processing water colour data from satellites and aircraft (see Chapter 5).

Using the measured volume scattering function $\beta(\theta)$ from Petzold (1972), Kirk

(1984) suggested via Monte Carlo simulations for turbid inland and coastal waters the following relationship between diffuse attenuation coefficient K_d and water colour producing agents (CPAs) (suppressing the spectral dependence here):

$$K_d(z_m, \theta_0') = (1/\mu_0)[a^2 + (0.473\mu_0 - 0.218)ab]^{1/2}. \qquad (1.26)$$

He had estimated a *single scattering albedo* $\omega_0 = b/(a+b) \leq 0.968$, equivalent to $b/a \leq 30$. This relationship incorporates solar zenith angle dependence via μ_0. z_m is the depth of the mid-point of the photic zone, which is the water column above $0.1E_d(\lambda, -0, \theta_0')$.

For overcast conditions a simplified version holds:

$$k_d(z_m) = 1.168[a^2 + 0.168ab]^{1/2}. \qquad (1.27)$$

Further on, $K_d(z_m, \theta_0')$ and $K_d(z_m)$ in eqs (1.26) and (1.27) will be referred to as $K_{d,sun}$ and $K_{d,sky}$.

For an incident irradiance with a diffuse fraction $F(F = E_{sky}/E_{sky} + E_{sun})$, transformation into F_w just below the water surface gives:

$$F_w = F(1 - \rho_{sky})/[F(1 - \rho_{sky}) + (1 - F)(1 - \rho_{sun}(\theta_0))]. \qquad (1.28)$$

Therefore, the resultant downwelling irradiance attenuation coefficient can be given as (Bukata *et al.*, 1995):

$$K(\lambda, \theta_0', -0) = F_w K_{sky}(\lambda) + (1 - F_w)K_{sun}(\lambda, \theta_0'). \qquad (1.29)$$

These expressions will be useful in our further discussions in Chapters 4 and 5.

1.4 TRANS-SPECTRAL PHOTON INTERACTIONS IN NATURAL WATERS

We have considered hitherto only the effects of *elastic* scattering (i.e. the scattered light has the same wavelength as the impinging one) and absorption on the water leaving radiance, and hence the water colour. However, Raman (inelastic) scattering by water molecules, fluorescence by both chlorophyllous pigments in phytoplankton, and dissolved organic molecules can also affect appreciably the spectral distribution of upwelling radiance, and consequently water colour (see, for example, Sugihara *et al.*, 1984; Fischer and Kronfeld, 1990; Vodacek *et al.*, 1994; Waters, 1995; Coble and Brophy, 1996).

This evidence stimulated the development of models simulating the above processes (Gordon, 1979; Marshall and Smith, 1990; Haltrin and Kattawar, 1993; Vodacek *et al.*, 1994; Sathyendranath and Platt, 1998) that are often called transspectral since stimulation at wavelength λ_1 leads to emissions at λ_2 with $\lambda_2 > \lambda_1$.

In a single scattering approximation, the upwelling light fields originating from various sources of different nature can be treated as independent (Gordon, 1979), and the bulk apparent optical property, $R(\lambda, z)$, can therefore be considered as the sum of the individual AOPs arising from each kind of sunlight interaction within the water column, such as absorption and elastic scattering in an optically semi-infinite aquatic medium, R_{es} (denoted herein below also as R_∞, see, for example, eq. (1.36)), inelastic (Raman) scattering R_r, chlorophyll fluorescence R_{chl}^f, and fluorescence of dissolved organic molecules R_{doc}^f.

In reality, when also considering trans-spectral processes such as Raman scattering or fluorescence, whose phase functions are nearly isotropic, multiple scattering events may need to be included. However, Sathyendranath and Platt (1998) have shown that, typically, the contribution by photons scattered twice or three times to water Raman scattering would only be about 10% of single scattering. The contribution of even higher orders, the fourth and fifth, would again only be of the order of 10% of scattering orders two and three. It is not unreasonable to assume that this should also hold for fluorescence by chlorophyll and dissolved organic matter (Sathyendranath and Platt, 1998).

1.4.1   Raman scattering by water

Raman scattering is a type of scattering of electromagnetic radiation in which light changes its frequency (as well as its phase) as it passes through a medium. Hence, it is called *inelastic* scattering as opposed to *elastic* scattering of light such as Rayleigh and Mie light scattering occurring in gaseous and liquid media without a wavelength or frequency shift.

This change in scattered light frequency or wavelength is due to light interaction with vibrational modes of the molecules constituting the medium. In water the dominant Raman line is generated by the modulation of the amplitude of the propagating light wave/vibration by the fundamental O–H stretching mode of the H_2O molecule.

The Raman band is relatively broad with a maximum at a wavenumber shift of $3330\,\text{cm}^{-1}$ (Waters, 1995). Since the differential Raman scattering cross-section of water molecules $(d\sigma_{H_2O}/d\Omega)_{90°} = 8.3 \times 10^{-30}\,\text{cm}^2\,\text{molecule}^{-1}\,\text{sr}^{-1}$ for the exiting wavelength $\lambda_{ex} = 488\,\text{nm}$ in a direction perpendicular to the polarization plane of the incoming light beam, when integrated over the Raman emission linewidth, one has to assess the possible contribution of the water Raman scattering to total upwelling radiance, L_u.

Recently, Gordon (1999) re-examined the contribution of Raman scattering to water leaving radiance, L_u, for chlorophyll concentrations $\leqslant 1\,\text{mg/m}^3$ and found that it is $\geqslant 8\%$ at wavelengths of interest for ocean colour remote sensing. Therefore, it cannot be ignored in ocean colour modelling and inference of water quality parameters. Similar conclusions were later reported by Loisel and Stramski (2000) and Boynton and Gordon (2000).

Integration of the differential Raman cross-section of water over the full solid angle (4π steradian), using eq. (1.30):

$$d\sigma_{H_2O}/d\Omega = (d\sigma_{H_2O}/d\Omega)_{90°} \frac{[\rho + (1 - \rho)\sin^2\alpha]}{(1 + \rho)} \tag{1.30}$$

(α = scattering angle, ρ = depolarization ratio of water) results in a volume scattering coefficient b_r,

$$b_r(\delta\nu_r) = N \int_{4\pi} \left(\frac{d\sigma_{H_2O}}{d\Omega}\right) d\Omega, \qquad (1.31)$$

where $\delta\nu_r$ = the Raman frequency shift and N = number of water molecules per cm^{-3} (at normal sea level pressure $N \cong 3.3 \times 10^{22}$ cm^{-3} (Fadeev et al., 1982)).

The integration of eq. (1.30) over 4π and the further integration over the Raman frequency shift distribution $\delta\nu_r$ results in $b_r = 2.6 \times 10^{-4}$ m^{-1} at $\lambda_{ex} = 488$ nm.

The wavelength-dependence of the Raman band and thus b_r is very strong and is proportional to λ_{em}^{-5}, whereby λ_{em} is the wavelength of the inelastically scattered light (Sugihara et al., 1984). Consequently, in order to calculate b_r at wavelengths other than 488 nm, a simple relationship can be used:

$$(d\sigma_{H_2O}/d\Omega)_{\lambda_i} = (d\sigma_{H_2O}/d\Omega)_{\lambda=488\,nm} \left(\frac{\nu_i - \overline{\delta\nu_r}}{\nu_{488} - \overline{\delta\nu_r}}\right)^5, \qquad (1.32)$$

where $\nu_i = 1/\lambda_i$ and $\overline{\delta\nu_r}$ is the mean Raman frequency shift.

Sugihara et al. (1984) report an expression for the Raman band maximum wavelength, $\lambda_{em\,max}$ (nm), as a function of the exiting wavelength (λ_{ex}, nm):

$$\lambda_{em\,max}(\lambda_{ex}) = \frac{\lambda_{ex}}{-3.357 \times 10^{-4}\lambda_{ex} + 1}. \qquad (1.33)$$

The Raman scattering phase function $\beta_r(\theta)$ can be approximated via (Porto, 1966):

$$\beta_r(\theta) = \frac{3}{16\pi}\frac{1+3\rho}{1+2\rho}\left(1 + \frac{1-\rho}{1+3\rho}\cos^2\theta\right), \qquad (1.34)$$

where ρ = depolarization ratio of water ($\rho = 0.17$ at $\overline{\delta\nu_r} = 3400$ cm^{-1} (Ge et al., 1995)).

Therefore, the Raman scattering function is symmetric and follows a distribution which is very similar to the phase function of elastic scattering by water (Morel and Gentili, 1991):

$$\beta_w(\theta) = \frac{3}{4\pi} \cdot \frac{3}{3+p}(1 + p\cos^2\theta), \qquad (1.35)$$

where the polarization factor $p = 0.84$ for pure water.

The value of $b_r(\lambda)$ has been determined by many workers (for references, see Barlett et al., 1998). It lies in the range $(2.7 \pm 0.2) \times 10^{-4}$ m^{-1} for $\lambda = 488$ nm. Desiderio (2000) points out that for calculations, in which the intensities are expressed in terms of quanta (as it generally occurs in Monte Carlo calculations that tract photon trajectories), the quantity $b_{r,q}(\lambda)$ should be used instead, obtained by multiplying $b_r(\lambda)$ with the ratio of the energy per incident quantum to the energy per scattered quantum. As pointed out above, the spectral dependence of $b_r(\lambda)$ is proportional to λ^{-5} (although it is slightly different for the incident and the Raman scattering wavelengths). The backscattering probability for elastic and inelastic scattering by water molecules is generally assumed to be 0.5 (Marshall and Smith, 1990) owing to the symmetric shape of the phase functions.

Barlett (1997) quantitatively compared four models (Marshall and Smith, 1990; Lee et al., 1994; Haltrin and Kattawar, 1993; Sathyendranath and Platt, 1998) calculating the Raman scattering contribution to the upwelling irradiance in natural waters. This comparison has shown that the model by Sathyendranath and Platt (1998), being most easily applicable, gives adequate results. Therefore, the volume reflectance just below the surface caused by Raman scattering can be assessed via:

$$R_r(\lambda_{em}, -0) = \frac{b_{b_r}(\lambda_{ex})E_d(\lambda_{ex}, -0)}{2\mu_d(K_u(\lambda_{em}) + K_d(\lambda_{ex}))E_d(\lambda_{em}, -0)}\left[1 + \frac{\mu_d}{\mu_u}R_\infty(\lambda_{ex}, -0) + \frac{b_b(\lambda_{em})}{2\mu_u K_u(\lambda_{em})}\right]$$

(1.36)

whereby the dependence on in-water refracted solar zenith angle θ_0' is omitted in eq. (1.36). $b_{b_r}(\lambda)$ is the Raman coefficient for backscattering; $b_{b_r}(\lambda) = B_r b_r(\lambda)$; B_r = inelastic backscattering probability; $b_r(\lambda)$ = Raman scattering coefficient; $b_b(\lambda)$ = elastic backscattering coefficient; $K_d(\lambda)$, $K_u(\lambda)$ = irradiance attenuation coefficients for downwelling and upwelling fluxes, respectively; μ_d, μ_u = average cosines for downwelling ($E_d(\lambda)$) and upwelling ($E_u(\lambda)$) irradiances, respectively; λ_{ex} and λ_{em} = wavelengths of excitation and emission of Raman scattering, respectively; and R_∞ is the water volume reflectance for an optically semi-infinite water medium.

In deriving eq. (1.36), it was assumed that the water body is vertically homogeneous, optically thick and not excessively absorption-dominated (see Barlett (1996) for details).

Expressions derived by Kirk (1984) can be used to obtain the diffuse attenuation coefficients for direct incident solar radiation, $K_{d,sun}(\lambda, \theta_0')$, for diffuse incident sky radiation, $K_{d,sky}(\lambda)$, and the combined downwelling diffuse irradiance attenuation coefficient (see eqs (1.26)–(1.29)).

As seen from eq. (1.36), determination of $R_r(\lambda_{cm}, -0)$ requires among other parameters, the multifunctional variable $E_d(\lambda, -0, \theta_0')$. For a calm water surface, the downwelling subsurface irradiance $E_d(\lambda, -0, \theta_0')$ can be directly estimated from above-surface values of downwelling irradiance $E_d(\lambda, +0, \theta_0)$ through the following relationship (Baker and Smith, 1990):

$$E_d(\lambda, -0, \theta_0) = \frac{1}{1 - \rho_u(\theta_0)}[(1 - \rho_{sun}(\theta_0))E_{sun}(\lambda, \theta_0) + (1 - \rho_{sky})E_{sky}(\lambda)], \quad (1.37)$$

where $\rho_u(\theta_0)$ is the reflection coefficient for subsurface upwelling irradiance for a solar zenith angle θ_0.

1.4.2   Fluorescence by phytoplankton and dissolved organic matter

In the world oceans, photosynthetic pigments of phytoplankton, first of all chlorophyll-a, and dissolved organic matter are natural fluorophores that can emit light under the impact of solar radiation propagating through the water column.

Gordon (1979) modified the radiative transfer equation to include the effect of fluorescent substances (phytoplankton chlorophyll-*a* as an example) and solved it in the single scattering approximation for a vertically homogeneous ocean containing fluorescent particles, with wavelength-independent quantum efficiency (η) and a Gaussian-shaped emission line centred at $\lambda = \lambda_{0\,em}$ with a width $\Delta\lambda_f$ at half-maximum of about 25 nm. For the wavelength $\lambda_{0\,em}$, the volume reflectance R^f arising from fluorescence by an aquatic constituent can be expressed for a given sun zenith angle, θ_0, as follows:

$$R^f(\lambda_{em}) = \frac{\eta}{\sqrt{2\pi\sigma^2}} \exp\left(-\frac{(\lambda_{em} - \lambda_{0\,em})^2}{2\sigma^2}\right) \Big/ 2K_d(\lambda_{em})\mu_0 E_d(\lambda_{em}, -0)\lambda_{em}$$

$$\times \int_{\lambda_{ex}} \lambda_{ex} a_f(\lambda_{ex}) E_d(\lambda_{ex}, -0) f(\lambda_{ex}, \lambda_{em}) \, d\lambda_{ex}, \tag{1.38}$$

where λ_{ex} and λ_{em} are the excitation and emission wavelengths, respectively, $a_f(\lambda)$ is the fluorophore absorption coefficient, $\sigma = \Delta_f/2.35 \approx 10.6$ nm, and

$$f(\lambda_{ex}, \lambda_{em}) = \frac{K_d(\lambda_{em})}{K_d(\lambda_{ex})}\left[1 + \frac{K_d(\lambda_{em})}{K_d(\lambda_{ex})} \ln\left(1 + \frac{K_d(\lambda_{ex})}{K_d(\lambda_{em})}\right)\right].$$

Vodacek *et al.* (1994) suggested a model of the sun-stimulated fluorescence of chromophoric dissolved organic matter (*cdom*). This model was extended by Green and Blough (1994) by characteristic values and spectral variations of the dissolved organic carbon (*doc*) fluorescence quantum yield, $\eta_{doc}(\lambda)$, and the absorption coefficient, $a_{doc}(\lambda)$. The model also provided a conversion of the relative fluorescence to absolute photon numbers relating the quantum yield η of *cdom* to that of quinine sulfate. Unlike for the above chlorophyll fluorescence model, it used an experimentally determined shape factor, otherwise being analogous to the formalism of Gordon (1979).

A few years later, considering the fluorescence emitted from a small volume of a water column, Culver and Perry (1997) suggested the following model for the fluorescence-related upwelling irradiance E_u^f caused by dissolved organic matter fluorescence at a depth z in a vertically homogeneous water column:

$$E_u^f(\lambda_{em}, z) = \tfrac{1}{2}\Psi(\lambda) \int_{\lambda_1}^{\lambda_2} \frac{E_{d_0}(\lambda_{ex}, z)}{K_d(\lambda_{ex}) + K_u^f(\lambda_{em})} \, d\lambda_{ex}, \tag{1.39}$$

where $\Psi(\lambda) = \eta_f(\lambda_{ex}) S_f(\lambda_{em}) \int_{\lambda_1}^{\lambda_2} a_f(\lambda_{ex}) \, d\lambda_{ex}$, η_f = fluorescence quantum yield, S_f = fluorescence band shape factor, a_f = absorption coefficient of the fluorophore, $E_{d_0}(\lambda)$ = downwelling scalar irradiance, and K_u^f = diffuse attenuation coefficient for the upwelling fluorescent flux.

Following Gordon (1979), Culver and Perry (1997) assume that in the case of phytoplankton the shape factor S_f can be approximated by a Gaussian distribution. Furthermore, based on considerations of the angular distribution of the subsurface light field (isotropic for fluorescence and inverse to the mean cosine for solar downwelling irradiance), Culver and Perry suggest that, just below the surface for $z = -0$ the following relationships hold: $K_u^f(\lambda_{em}) = \mu_0 K_d(\lambda_{em})$ and $E_{d_0}(\lambda) = E_d(\lambda)/\mu_0$. With these assumptions in mind, the volume reflectance component originating from the fluorescence of an aquatic constituent can be expressed as:

$$R^f(\lambda_{em}) = \frac{1}{2\sqrt{2\pi\sigma^2}E_d(\lambda_{em}, -0)} \exp\left(-\frac{(\lambda_{em} - \lambda_{0em})^2}{2\sigma^2}\right)$$

$$\times \int_{\lambda_{ex}} \eta_f(\lambda_{ex}) a_f(\lambda_{ex}) \frac{E_d(\lambda_{ex}, -0)}{\mu_0(K_d(\lambda_{ex}) + 2\mu_0 K_d(\lambda_{em}))} d\lambda_{ex}. \qquad (1.40)$$

By comparing (1.38) and (1.40), it can be seen that both expressions would become very similar if the limitation of Gordon's model (wavelength-independence of η) is lifted.

1.5   IMPACT OF BOTTOM REFLECTION ON UPWELLING RADIANCE AND VOLUME REFLECTANCE IN WATER

In addition to absorption, elastic scattering, water Raman (inelastic) scattering, fluorescence by phytoplankton chlorophyll and dissolved organics, bottom reflection or *albedo* (defined as the ratio of irradiance reflected to the incoming irradiance) affects the water colour in optically shallow waters. Following Estep, and some other workers (for references, see Kondratyev *et al.*, 1999), Tolk *et al.* (2000) have emphasized the importance of spectral bottom reflection impact on water colour.

In optically shallow waters a certain number of photons propagating downwards can be reflected at the bottom. Then the upwelling irradiance $E_{u,tot}(\lambda, -0)$ is the sum of two components, $E_{u,wc}(\lambda, -0)$ and $E_{u,bot}(\lambda, -0)$:

$$E_{u,tot}(\lambda, -0) = E_{u,wc}(\lambda, -0) + E_{u,bot}(\lambda, -0), \qquad (1.41)$$

where $E_{u,wc}(\lambda, -0)$ and $E_{u,bot}(\lambda, -0)$ are the upwelling irradiances due to photons which have not interacted or have interacted with the bottom, respectively.

Thus, the resulting diffuse volume reflectance just beneath the water surface $R_{tot}(\lambda, -0)$ can be decomposed as well:

$$R_{tot}(\lambda, -0) = R_w(\lambda, -0) + R_{bot}(\lambda, -0), \qquad (1.42)$$

where $R_w(\lambda, -0)$ is the spectral volume reflectance due to photons which were scattered back to the water surface without interaction with the bottom (this entity can equally be denoted as R_∞), and $R_{bot}(\lambda, -0)$ is the spectral reflectance arising from photons reflected or backscattered at the bottom.

Assuming that the bottom is a Lambertian reflector with an albedo A, the reflected irradiance that reaches the water–air interface ($z = -0$) can be defined as:

$$E_{u,bot}(\lambda, -0, H) = A(\lambda)E_d(\lambda, -0) \exp[-(K_d(\lambda) + K_u(\lambda))H], \qquad (1.43)$$

where H is the bottom depth, and K_d and K_u, as above, are the downwelling and upwelling diffuse irradiance attenuation coefficients.

It can be easily shown that eq. (1.41) then can be rewritten into:

$$E_{u,tot}(\lambda, -0) = E_d(\lambda, -0)\{R_w(\lambda, -0)[1 - \exp(-(K_d(\lambda) + K_u(\lambda)H)]$$
$$+ A(\lambda)\exp[-(K_d(\lambda) + K_u(\lambda))H]\}. \tag{1.44}$$

If one assumes that K_u equals K_d (a rough approximation), then both can be replaced by an operational and unique K coefficient, and eq. (1.44) becomes (the wavelength-dependence is omitted):

$$E_{u,tot}(-0) = E_d(-0)\{R_w[1 - \exp(-2KH)] + A\exp(-2KH)\}$$
$$= E_d(-0)[R_w - (R_w - A)\exp(-2KH)]. \tag{1.45}$$

Now, the total diffuse volume reflectance $R_{tot}(\lambda, -0)$ can thus be simplified to:

$$R_{tot} = R_w + (A - R_w)\exp(-2KH). \tag{1.46}$$

It is worth mentioning that the relationship (1.46) derived by Maritorena *et al.* (1994) is identical to the analytical expression obtained independently by Estep (1994).

2

Optical properties of non-Case I waters: hydro-optical models

2.1 COMPOSITION OF NATURAL WATER AND ITS RELATION TO HYDRO-OPTICAL ASPECTS OF THE PROBLEM

Natural waters are complex media comprising living, non-living, and once-living material that may be present in aqueous solution or suspension. Together with air bubbles and inhomogeneities resulting from small-scale water eddies, all these components determine the bulk optical properties of natural water bodies.

Pure water is defined as a *chemically* pure substance composed of a mixture of several water isotopes, each of different molecular mass.

The principal organisms present in water are *plankton*, a collective term encompassing all vegetable and animal organisms suspended in water (either hovering or floating), unable to resist the current, and not rigidly connected to the bottom. Plankton include animal organisms (*zooplankton*), algal plant organisms (*phytoplankton*), bacteria (*bacterioplankton*), and lower plant forms such as *algal fungi*.

These organisms represent the lowest levels in the food chain relationships involving not only the plankton (and their essential nutrients), but also higher forms of aquatic life. Phytoplankton consume nutrients (biogenes) from their surrounding waters and in the presence of subsurface sunlight synthesize these nutrients into organic matter through the process of *primary production*. Zooplankton graze on phytoplankton. As consequences of their vital functions and their mortality, zooplankton generate *secondary* organic matter.

The number of bacterioplankton reaches 10^5 cells per cm^3 in oligotrophic waters, a figure that becomes considerably larger as the water becomes more rich in nutrients, reaching about 10^9 cells per cm^3. The colourless part of bacterioplankton, of course, do not produce an impact on the aquatic colour, although they probably strongly contribute to scattering (see below). The coloured bacterioplankton possess pigments that are comparable to those of phytoplankton: the absorption spectra of the bacterial chlorophyll are similar to those of algal chlorophylls, except that the peaks are shifted towards longer wavelengths.

Algal fungi, along with bacterioplankton, are the sole transformers of organic matter in natural waters into compounds digestible by most aquatic microorganisms. It should be underscored that the complete lack of fungal coloration, coupled with their low concentrations (Kondratyev et al., 1999), results in algal fungi producing no optical impact on natural waters.

The concerted action of algal fungi and bacterioplankton in the decomposition of the available aquatic organic matter in suspension determines the dynamics of the food chain at the lowest trophic level. Owing to the low-level food chain dynamics, dissolved organic matter (*dom*) (termed the *autochtonous dom* component) is indigenous to natural waters. Lakes and marine/oceanic *coastal* waters are also receptacles of *dom* inputs from surface runoff and river discharges which contribute an *allochthonic dom* component. These runoff and river discharge inputs also introduce a suspended inorganic matter (*sm*) component to non-Case I waters.

Phytoplankton cells and colonies exist in a variety of shapes (filaments, ribbons, stars, etc.) and sizes. Reynolds (1984) and Petrova (1990) present statistical data on the phytoplankton sizes, which show that unicellular algae possess nominal mean maximum dimensions in the range 5 to 40 µm. The corresponding range for filaments would be 80 to 150 µm, for ribbon colonies 60 to several hundred µm, and for mucilaginous colonies 50 to 200 µm.

The coloration of phytoplankton cells is dependent on their pigment content and composition. Phytoplankton pigments serve as the collectors and suppliers of energy for the photosynthesis, the basic mechanism in plant growth. There are three basic types of photosynthesising pigments: *chlorophylls*, *carotenoids*, and *phycobilins*. Chlorophylls and carotenoids are present in all algal species. Phycobilins are additionally present in blue-green algae and dinoflagellates. The principal chlorophyll absorption occurs in the short wavelength (blue) and, generally to a lesser degree, the long wavelength (red) regions in the visible spectrum. Carotenoid absorption peaks in the blue spectral region. Phycobilin absorption strongly depends upon species with peaks at green, yellow, or short red wavelengths.

Zooplankton, as well as insects and higher plants, serve as food for fish that, in turn, extend the food chain to aquatic and terrestrial carnivores and herbivores. As such, zooplankton are in dynamic equilibrium with the other components of the aquatic food chain. Depending upon the trophic status of non-Case I waters, the concentrations of zooplankton can easily exceed several hundreds of thousands of plankters per m^3, varying in size from 30 µm to >2 mm. Since zooplankton are grazers of phytoplankton and phytoplankton are chlorophyll-bearing biota, chlorophyll is invariably present within the digestive tracts of zooplankters. Thus this presence of chlorophyll and its derivatives would suggest that absorption and scattering of zooplankton might display features comparable to the counter-characteristics of phytoplankton. However, we are not aware of data that would support or refute this hypothesis. Besides, the zooplankter concentration is generally low (compared to bacterioplankton and phytoplankton, see below), and on this basis, zooplankton are tentatively considered to be ineffective for light transfer through the water body, although this assumption does not necessarily hold for all cases, it does at least in lacustrine waters (Petrova, 1990).

Inorganic salts dissolved in natural waters affect absorption and scattering within the water column. However, their most significant absorption lies at the ultraviolet, rather than visible, wavelengths since the electron transition bands of dissolved inorganic salts exist at $\lambda < 300$ nm (Shifrin, 1983). The contribution of scattering by dissolved inorganic salts accounts for about 20% to 30% of total scattering in offshore/pelagic oceanic waters of average salinity of 35 $\mathrm{kg\,m^{-3}}$. In inland fresh water bodies, as well as brackish coastal waters (diluted with fresh water arriving as river runoff as well as with flows of rain and snowmelt accumulations from the catchment), the contribution to scattering by dissolved inorganic salts in non-Case I water is strongly reduced.

Dissolved organic matter (*dom*) concentrations are the consequence of either photosynthetic activity of phytoplankton (autochthonic) or direct inputs of terrestrial matter (allochthonic). Of the total organic matter resulting from phytoplankton photosynthesis, up to 20% can be released to the ambient aquatic environment through *metabolic egestion* (Gorlenko *et al.*, 1974). These egesta are, in general, assimilated by bacterioplankton; however, in nutrient-limited waters, the phytoplankton themselves may compete with bacterioplankton for such egesta assimilation.

In fresh and brackish waters *dom* is not conservative, but rather undergoes various biological and *photochemical* transformations (Bertilsson and Travnik, 2000) before the residual of *dom* is eventually discharged into the oceans. Solar radiation, particularly in the ultraviolet, has the potential (especially in oligotrophic rather than eutrophic non-Case I waters) to alter the spectral and molecular properties of *dom* (mostly allochthonous *dom*), and to promote its degradation, either directly through *photooxidation* or indirectly by increased *bioavailability*.

Yellowness/brownishness is a characteristic hue of waters containing *dom* in large amounts, which is due to yellow and brown melanoids. These melanoids, however, comprise only a fraction ranging between 10% and 40% of the aquatic *dom*.

Since allochthonic *dom* is absent from mid-oceanic (Case I) waters, and the autochthonic *dom* should be low because of generally low nutrient levels in Case I waters, the aquatic humus in Case I waters co-varies with phytoplankton (or its proxy, chlorophyll), keeping in mind that the mid-oceanic waters are also nearly devoid of suspended minerals. As will be discussed in further chapters, this facilitates algorithm-development for remote sensing of Case I waters that is not available for remote sensing of inland and coastal waters.

Aquatic humus, *dom*, concentrations in oceanic waters are generally very low in the range 0.001–0.005 g Carbon per $\mathrm{m^3}$. However, it should be pointed out that there are observations (Church *et al.*, 2002) indicating that a small fraction of the annually produced organic matter in some locations of the world oceans can escape biological utilization on time scales of months to years. A conjecture has been put forward that it might be a result of reorganization of dynamics of plankton community, which brings about significant alterations to cycling of organic matter in the ecosystem. It is believed that can be driven by basin-scale climate variability.

The concentration or density of *dom* is usually given in the unit grams of *carbon* per unit volume since carbon constitutes approximately *half* the *dom* by weight. Because of this, *dom* will here and henceforth be named *doc* (where *c* stands for carbon). The concentration of *doc* in oceanic waters is about four orders of magnitude below the concentrations of dissolved salts. However, the optical impact of *doc* is stronger than that of dissolved salts. Although *doc* does not significantly impact scattering within the water column, it considerably increases absorption. In inland and coastal waters, however, *doc* concentrations are often considerably higher: *doc* ranges from a 1 to $2\,\text{g}\,\text{C}\,\text{per}\,\text{m}^3$ in oligo/mesotrophic waters to $20\text{--}25\,\text{g}\,\text{C}\,\text{per}\,\text{m}^3$ in hyper-eutrophic waters (Romankevich, 1977).

The absorption by *doc* decreases exponentially from short to long wavelengths in the visible spectrum. The observed *doc* absorption in the visible spectrum is a 'tail' of electron transition absorption bands located in the ultraviolet (UV) part of the spectrum. The spectral dependence of the absorption coefficient $a_{ys}(\lambda)$ for the coloured fraction of *doc* (so-called *yellow* substance), can be approximated well by a simple exponential function (Zepp and Schlotzhauer, 1981):

$$a_{ys}(\lambda) = a_{ys}(\lambda_0) \exp[-s(\lambda - \lambda_0)], \qquad (2.1)$$

where *s* is a slope parameter that is assumed to be independent of wavelength λ. Thus the constancy of the shape of the $a_{ys}(\lambda)$ versus λ curve in eq. (2.1) enables the selection of a reference wavelength, λ_0, upon which to compare the yellow substance concentrations of different waters. The selection of $\lambda_0 = 400\,\text{nm}$ is certainly appropriate since it represents the maximum a_{ys} in the visible spectrum. However, $\lambda_0 = 440\,\text{nm}$ is also frequently encountered in the published literature.

To date, little attention has been directed towards the possible effects of *doc* on the scattering properties of natural waters (we, too, in our analyses of volume reflectance/water quality relationships in non-Case I waters find it convenient to set $b_{doc}(\lambda)$ to zero). Although such neglect of the scattering from organic matter that resides in the water column is justified for true molecular solutions, there is sometimes a very substantial fraction of 'dissolved' organic matter that is not in true molecular solution, but exists in colloidal form, which can impact the bulk scattering coefficient characterizing the optical properties of the water column.

More importantly, a certain, presumably small, fraction of *doc* can fluoresce with the emission maximum(s) located in a wide spectral region extending from about 310 to 520 nm depending on the nature of *doc* and the excitation wavelength (Coble, 1994).

Suspended matter. All natural water bodies contain suspended matter consisting of organic and inorganic matter under the collective term *seston*. Seston is extremely diverse in origin and in composition, and includes mineral terrigenous particles, plankton (see above), detritus (largely residual products of the decomposition of phytoplankton and zooplankton cells as well as macrophytic plants), volcanic ash particles, particulates resulting from *in situ* chemical reactions, and particles of anthropogenic origin.

Suspended inorganic (mineral) matter (sm). The presence of terrigenous suspended particles is a consequence of river discharge, coastal erosion, catchment runoff, long- and short-range transport of atmospheric particulates followed by dry and wet deposition.

The main constituents found in *sm* derived from surface soils are quartz, feldspars, calcite, dolomite, gypsum, mica, kaolinite, illite, montmorillonite, palygorskite, chlorite (for references see Sokolik and Toon, 1999). The real and imaginary parts of the complex refractive index for the atmospherically transported *sm* vary at $\lambda = 500$ nm between 1.4 and about 2.8 (and even 3.4 for haematite), and <0.000 05 to typically 0.001 (for haematite up to 0.1), respectively. The *single scattering albedo*, $\omega = b/c$, at $\lambda = 500$ nm is between 0.95 and 1.0 (it decreases down to 0.8 at $\lambda \leq 300$ nm, but it is about 0.6 over large parts of the visible spectrum for haematite). The *asymmetry parameter* (b_f/b) is generally found between 0.68 and 0.89 (Sokolik and Toon, 1999). According to Kondratyev *et al.* (1999), fine clay particulates rarely exceed 3 to 4 μm in diameter, silt particles are in the range 5 to 40 μm, very fine grain sands fall in the range 40 to 130 μm, and coarse grain sands are in the range 130 to 250 μm.

In Lake Ladoga, seston concentrations range from 0.1 to 12.0 g per m^3, with suspended minerals accounting for about 50% to 97% (Multi-author, 1987). Suspended mineral concentrations in Lake Ontario have been reported in the range 0.2 to 8.9 g per m^3 (Bukata *et al.* 1995).

Along with the plankton, inorganic suspended particulates dictate the manner in which the subsurface light field is distributed in the natural water body and, therefore, have to be considered in terms of both their absorption and scattering properties. About 90% of the coarse fraction of the terrigenous matter is found within the nearshore zone.

Suspended organic matter. Suspended organics in natural water bodies contain planktonic organisms and detritus particles. As pointed out earlier, the plankton consist of algal fungi, bacteria (bacterioplankton), micro-algae (phytoplankton) and micro-animal organisms (zooplankton). With the only exception of algal fungi (which are transparent and colourless and thus have a refraction index practically indistinguishable from that of water), all planktonic species are optically active in the visible spectrum as a result of

(1) light absorption, mostly, stemming from pigments either indigenous to the aquatic cells (bacterioplankton and phytoplankton) or otherwise acquired through trophic chain interactions (zooplankton) (Karnaukhov, 1988);
(2) light scattering by cells/bodies, which are generally of complex and very irregular geometry and inner structure. It is due to their small size that the viruses and bacteria have the highest backscattering ratio (defined as the ratio of backscattering to the total scattering) among autotrophic and heterotrophic plankton (van de Hulst, 1957).

When deciding on the eligibility of water co-existing constituents for the status of principal CPAs that dictate the bulk water optical/colour properties, their absolute and relative concentrations become the criterion. The Lake Ladoga water can be considered as a good example. In spring, the bacterioplankton number concentration (C_{bp}) in the coastal zone may attain 1.5–2.0 million cells cm^{-3}, whereas within the pelagic region, C_{bp} does not generally exceed 0.1 to 1.0 million cells cm^{-3}.

In summer, C_{bp} decreases in coastal waters to about 0.5 to 0.65 million cells cm^{-3} but increases up to 10 to 20 million cm^{-3} in the central part of the lake. In autumn, C_{bp} is about 0.6 to 0.7 million cm^{-3} and is nearly uniformly distributed throughout the lake (Viljanem et al., 1996).

The phytoplankton concentration (C_{php}) within the top (0–10 m) layer varies significantly depending on the vegetation period. The average C_{php} value might reach the level of several hundred thousand cells per litre in the pelagic part of the lake and several million cells per litre in the littoral zone (Viljanen et al., 1996). The typical concentrations of zooplankton (C_{zp}) in the surface waters (0–20 m) of the littoral zone are several hundreds of plankters per cubic metre. Generally, C_{zp} is, however, somewhat lower in the pelagic zone (Viljanen et al., 1996).

Consequently, taking the above numbers, the concentration ratios of bacterio-plankton/phytoplankton/zooplankton for maximal and minimal values of are as follows: $10^{12}/10^6/10^2$ and $10^{11}/10^5/10^1$. Although, as was emphasized earlier in this section, in some vegetation periods (of rather short duration), the concentration of phytoplankton and zooplankton could be of the same order of magnitude in inland water bodies (Petrova, 1990; Viljanen et al., 1996) it can, nevertheless, be assumed that zooplankton are hardly to be generally considered as a major colour producing agent (CPA).

Detritus. Detritus, as defined above, consists of suspended organic and partly mineralized particulates which are fragments of dead plankton and faecal pellets of zooplankton. The detritus content in water can vary substantially depending on vegetation period and trophic level. The detritus concentration is closely related to the phytoplankton content, and it reaches 10% to 30% of C_{php}. Although these concentration ratios are typical of Lake Ladoga, they are also valid for other temperate inland fresh water bodies (Petrova, 1990).

Air bubbles. In the upper layers of natural waters trapped air bubbles are mainly generated by injection of air by breaking waves. The refractive index of air bubbles per se is 0.75. The majority of the air bubbles injected into the surface layers of natural waters are unstable. However, for oceanic air bubbles with long residence times, a bubble concentration of about 10^5 to 10^6 m^{-3}, and even as large as 2.13×10^7, was observed (for references see Yan et al., 2002). The sizes (radius, r) of air bubbles vary from about 0.01 to about 350 μm, following on average the bubble size distribution $n(r) \sim r^4$ (Stramski, 1994). The bubbles are distributed within the top layer of the ocean, and observations indicate that the thickness of this layer can vary over a large range from 0.25 to 36 m. There is evidence (for references see Yan et al., 2002) that some bubble clouds could reach mean depths of about $4H_s$, where H_s is the significant wave height. But some bubble clouds can extend down to approximately $6H_s$: they are frequently observed at depths of 6 to 11 m and even down to 36 m (Wu, 1988). Given the bubble number density that has typically been reported from measurements in the sea (i.e. ranging, as we saw above, from about 10^5 to 10^7 cm^{-3}), the bubble population will significantly influence the scattering processes in the ocean, especially in oligotrophic waters. Little as yet is known about the dependence of bubble micro-physical properties on wind speed.

Often water bubbles are covered with organic films composed mainly of proteins or lipids. The thickness of such coatings of bubbles in sea water has been estimated to range from 0.01 μm for lipids to 1 μm for proteins. The mean relative refractive indices of the coating substances ($n \cong 1.20$ for proteins and $n \cong 1.10$ for lipids) are quite different from those of clean water bubbles (see above). As shown by Zhang *et al.* (1998), this implies that the coated bubbles scatter more strongly than clean bubbles. In any case, these air-bubbles-related scattering centres should also result in attenuation of downwelling radiation because of absorption, as the imaginary part of the refractive index of coated bubbles lies between 0.001 and 0.006, which is similar to the mean imaginary index of refraction of phytoplankton cells. The latter value is probably close to the maximum value for chlorophyll in the red absorption band (Morel, 1990). But also backscattered upwelling radiation is increased. Both effects change volume reflectance, which then may be erroneously attributed to the presence of suspended organic or inorganic particulates.

According to Zhang *et al.* (1998), the mean scattering efficiency factor of bubbles is comparable to those of nanoplankton (2–20) μm and microplankton (20–200) μm. However, for the *backscattering* ratio/efficiency, ranging from 0.02 for clean/ uncoated bubbles to as high as 0.08 for bubbles with a 0.1-μm thick protein cover, bubbles are at least one order of magnitude more efficient in backscattering than planktonic organisms. In general, the role of clean/coated bubbles in increasing/ modifying the resultant volume reflectance will decrease as the concentration of phytoplankton and suspended minerals increases. However, this decreasing contribution will shrink with increasing λ. This is due to b_b being spectrally flat for both clean and coated bubbles (Zhang *et al.*, 1998), whereas it decreases with λ for seston.

Therefore, air bubbles can generally be a reason for a nonzero upwelling radiance in the red and near-infrared part of the spectrum in phytoplankton-free deep oceans, thus invalidating the assumption of black water that lies at the basis of most atmospheric correction procedures (see Chapter 5).

The extreme diversity of the organic and inorganic components of inland and marine/oceanic waters, coupled with the temporal and spatial variability of the composition of these waters, presents a severe obstacle to defining the major players in water optics. Thus, a manageable number of these components (or surro- gates/proxies to these components) has to be selected for a physical model that adequately describes the optical properties of a natural water body.

At the same time, based on some *a priori* knowledge from molecular spectros- copy, some natural water constituents can be confidently ruled out from the list of those substances which we qualified in Chapter 1 as the main CPAs. For instance, *dissolved atmospheric gases* with the sole exception of oxygen do not absorb appre- ciably in the visible spectrum. Nor, at their small concentrations, can they contribute considerably to molecular cumulative scattering. Algal fungi and zooplankton are also likely be to excluded from the list of CPAs. Summarizing the above data, we suggest that the group of major CPAs for non-Case I waters should be composed of water per se, phytoplankton, bacterioplankton, *doc*, *sm*, detritus (in some cases), and probably air bubbles (in *oligotrophic* waters).

However, our literature search (Kondratyev *et al.*, 1999) indicates that, unfortunately, the data on microphysical characteristics and optical properties of bacterioplankton, detritus, and air bubbles are still scarce and do not yet allow any generalization. On the other hand, we believe that the exclusion of zooplankton, detritus, bacterioplankton, and air bubbles will not significantly detract from the discussion of the four-component model: water per se, phytoplankton, suspended minerals (*sm*) and dissolved organics (*doc*) in Case II waters.

We do feel, however, that the explicit inclusion of bacterioplankton, detritus, and air bubbles as well as the differentiation of individual planktonic taxa could greatly improve the capabilities of a bio-optical model. Such improvement should, of course, be considered in the future. At present, the harsh reality of the optical complex inland and coastal waters and the need to match the number of independent equations with the number of unknowns restrict the realistic options.

2.2   HYDRO-OPTICAL MODELS

2.2.1   Pure water

The absorption spectrum of pure water (Fig. 2.1) shows a deep minimum at the short and middle wavelengths of the visible, followed by a continuous increase in absorption at $\lambda > 500$ nm. The absorption in the red part of the spectrum is a peripheral manifestation of much more intense IR absorption bands located at $\lambda > 700$ nm. A pronounced increase of water absorption in the UV spectrum is due to electron transitions occurring within the water molecule.

The scattering properties of pure water are generally governed by the Rayleigh scattering mechanism, so that elastic scattering of light on fluctuations of water density displays a spectral dependency roughly proportional to $\lambda^{-4.3}$, whereas its angular distribution follows the well-known Einstein–Smoluchovsky equation (Whitlock *et al.*, 1981).

A detailed overview of experimental data on the optical characteristics of pure water may be found in Multi-author (1983). Numerical values of the absorption and scattering properties of pure water have been reported by Hulbert (1945), Tam and Patel (1979), Smith and Baker (1981), Buiteveld *et al.* (1994), Sogandares and Fry (1997), Pope and Fry (1997), and some others (Fig. 2.1).

The absorption and scattering coefficients listed in Table 2.1 refer to a water temperature $T = 20°C$. It should be noted that water absorption properties vary with temperature: the liquid water absorption coefficient a_w increases with water temperature T in the spectral region from 400 to 780 nm and it decreases in the spectral interval from 780 to 800 nm. In the temperature range $0.5°C \leq T \leq 26°C$, the variation of a_w is less than a few per cent. With the increasing temperature, the absorption maximum shifts to shorter wavelengths (Hakvoort, 1994). As seen, scattering by water molecules becomes insignificant when compared to absorption by

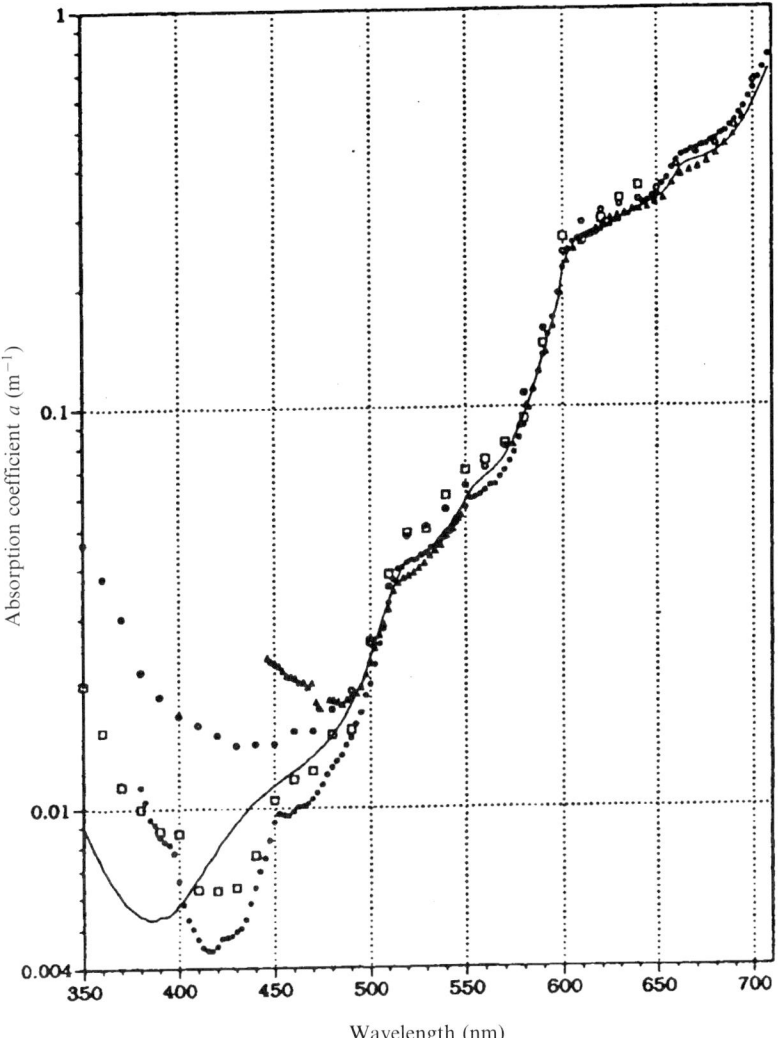

Figure 2.1. A comparison of results of the absorption of pure water as reported by different authors: ● (Pope and Fry, 1997), smooth curve (Buiteveld *et al.*, 1994), △ (Tam and Patel, 1979), ○ (Smith and Baker, 1981), □ (Sogandares and Fry, 1997).

water molecules for $\lambda > \sim 580$ nm. Then the attenuation of light (C_w) is essentially a consequence of molecular absorption (i.e. $C_w \approx a_w$). For λ in the range from 520 to 400 nm, however, scattering by water molecules (b_w) is either important or dominant, so that c_w is primarily dictated by b_w in the blue region of the spectrum.

Even a cursory analysis of the data listed in Table 2.1 indicates that there is a good deal of discrepancy in terms of spectral absorption values reported by different workers, especially at $\lambda \leq 500$ nm. Importantly, with the improvement of

Table 2.1. Absorption and scattering coefficients of pure water: (Smith and Baker, 1981)[SB], (Multi-author, 1983)[MA], (Buiteveld *et al.*, 1994)[BHD], (Pope and Fry, 1997)[PF].

λ, nm	n	$[a_w m^{-1}]^{SB}$	$[a_w, m^{-1}]^{MA}$	$[a_w, m^{-1}]^{BHD}$	$[a_w, m^{-1}]^{PF}$	$[b_w, m^{-1}]^{SB}$	$[b_w, m^{-1}]^{MA}$	$[b_w, m^{-1}]^{BHD}$
250	1.377	0.559	0.190			0.0443	0.032	
300	1.359	0.141	0.040	0.0174		0.0201	0.015	0.0174
320	1.354	0.0844	0.020	0.0133		0.0153	0.012	0.0133
350	1.349	0.0463	0.012	0.0092		0.0103	0.0082	0.0092
400	1.343	0.0171	0.006	0.0058	0.00663	0.0058	0.0108	0.0053
410		0.0162		0.0067	0.00473	0.0052		0.0048
420	1.342	0.0153	0.005	0.0079	0.00454	0.0047	0.0040	0.0043
430		0.0144		0.0092	0.00495	0.0042		0.0039
440	1.340	0.0145	0.004	0.0104	0.00635	0.0038	0.0032	0.0036
450		0.0145		0.0114	0.00922	0.0035		0.0033
460	1.339	0.0156	0.002	0.0124	0.00979	0.0031	0.0027	0.0030
470		0.0156		0.0135	0.0106	0.0029		0.0027
480	1.337	0.0176	0.003	0.0152	0.0127	0.0026	0.0022	0.0025
490		0.0196		0.0181	0.0150	0.0024		0.0023
500	1.336	0.0257	0.006	0.0238	0.0204	0.0022	0.0019	0.0021
510		0.0357		0.0329	0.0325	0.0020		0.0019
520	1.335	0.0477	0.014	0.0409	0.0409	0.0019	0.0016	0.0018
530	1.335	0.0507	0.022	0.0429	0.0434	0.0017	0.0015	0.0017
540	1.335	0.0558	0.029	0.0495	0.0474	0.0016	0.0014	0.0015
550	1.334	0.0638	0.035	0.0588	0.0565	0.0015	0.0013	0.0014
560	1.334	0.0708	0.039	0.0672	0.0619	0.0014	0.0012	0.0013
570		0.0799	0.0759		0.0695	0.0013		0.0012
580	1.333	0.108	0.074	0.0952	0.0896	0.0012	0.0011	0.0011
590		0.157		0.1356	0.1351	0.0011		0.0011
600	1.333	0.244	0.200	0.2224	0.2224	0.0011	0.00093	0.0010
610		0.289		0.2691	0.2644	0.0010		0.0009
620	1.332	0.309	0.240	0.2810	0.2755	0.0009	0.00082	0.0009
630		0.319		0.2955	0.2916	0.0009		0.0008
640	1.332	0.329	0.270	0.3111	0.3108	0.0008	0.00072	0.0008
650		0.349		0.3315	0.340	0.0007		0.0007
660	1.331	0.400	0.310	0.3791	0.410	0.0007	0.00064	0.0007
670		0.430		0.4122	0.439	0.0007		0.0006
680	1.331	0.450	0.380	0.4318	0.465	0.0006	0.00056	0.0006
690		0.500		0.4760	0.516	0.0006		0.0006
700	1.330	0.650	0.600	0.5722	0.624	0.0005	0.0050	0.0005
710		0.839		0.7415	0.827	0.0005		0.0005
720		1.169		1.0724	1.231	0.0005		0.0005
730		1.799		1.6211	1.778	0.0005		0.0004
740	1.329	2.38	2.250	2.5319		0.0004	0.00040	0.0004
750	1.329	2.47	2.620	2.7334		0.0004	0.00039	0.0004
760	1.329	2.55	2.560	2.7710		0.0004	0.00035	0.0004
770		2.51		2.7542		0.0004		0.0003
780		2.36		2.6590		0.0003		0.0003
790		2.16		2.4924		0.0003		0.0003
800	1.328	2.07	2.020	2.2932		0.0003	0.00029	0.0003

experimental techniques and methodological approaches, it appears that the reported absorption minimum in the spectral region 410–420 nm has deepened reaching very small values of about $0.004\,\text{m}^{-1}$. An obvious discrepancy in a_w variations at $\lambda < 400$ nm is the one reported by Buiteveld *et al.* (1994). It may be due to the fact that Buiteveld *et al.* (1994) assessed a_w by subtracting estimates of the scattering contribution from the total attenuation coefficients measured by Boivin *et al.* (1986). On the other hand, Pope and Fry (1997) conducted direct absorption measurements throughout the entire spectral region. Interestingly, as can be seen in Table 2.1, the $a_w(\lambda)$ values in this spectral region suggested in the reference (Multi-author, 1983) compare well with the corresponding data by Pope and Fry (1997).

At wavelengths >500 nm (where no measurements with 'super' accuracies are required because $a_w(\lambda)$ is sufficiently strong), all workers report very similar results. The same refers to the values of $b_w(\lambda)$ in the blue.

Despite the obvious uncertainties existing in the literature for $a_w(\lambda)$ for $\lambda < 500$ nm and accounting for the arguments in Pope and Fry (1997), it seems to us that the combination of the Pope and Fry data in the range from 380 to 700 nm and the Sogandares and Fry (1997) data for $\lambda < 380$ nm should provide the most adequate spectral distribution of $a_w(\lambda)$ values at present.

As to the choice of $b_w(\lambda)$ data, we propose to choose the $b_w(\lambda)$ spectral distribution given in the reference (Multi-author, 1983) as Pope and Fry (1997) did not report $b_w(\lambda)$ values. This choice is partially motivated by the fact that the $a_w(\lambda)$ values reported in both papers (see Table 2.1) are at least the closest to one another.

The backscattering probability $b_{b_w} = 0.5$ because, as stated above, the scattering properties of pure water are governed by the Rayleigh scattering mechanism, and hence the phase function is symmetrical.

Although the issue of accurate values for absorption and backscattering properties of pure water is important per se, it can be admitted, here, that it is more important for clear oceanic waters, and, to a lesser degree, for Case II waters. Indeed, the differences in $a_w(\lambda)$ values at *short wavelengths* between Smith and Baker (1981) and Pope and Fry (1997), leaving alone the differences between Pope and Fry (1997) and Multi-author (1983) are not really too important, because even in relatively clear (e.g. oligotrophic) Case II waters they have a chance to be 'drowned' in the total absorption or/and scattering characteristics of natural waters. However, in our numerical simulations we adhere to the choice specified above.

2.2.2 Suspended inorganic (mineral) matter (*sm*)

In coastal zones and inland water bodies, suspended sediment concentrations in surface waters can be fairly high. Not infrequently, a large portion of the *sm* coarse fraction is found not only in the littoral but also pelagic zones of a large lake. The chemical composition of *sm* indicates (see Section 2.1) that these particles are mainly composed of quartz, and also silica, clays and calcites. These materials have low imaginary parts of the complex index of refraction (10^{-3}–10^{-4}), and their

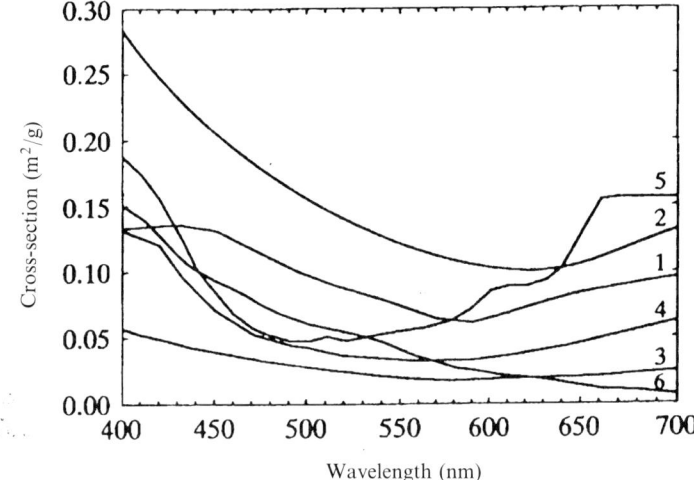

Figure 2.2. An intercomparison of independent determinations of the absorption cross-section spectra for non-chlorophyllous matter: 1, Bukata *et al.*, 1985; 2, Bukata *et al.*, 1981; 3, Gallie and Murtha, 1992; 4, Morel and Prieur, 1977; 5, Prieur and Sathyendranath, 1981; 6, Bowers *et al.*, 1996.

absorption increases towards the near UV spectral region as well as at $\lambda > 700$ nm (Kondratyev *et al.*, 1983).

There are only a few published values for the absorption and backscattering cross-section spectra of suspended minerals [$a_{sm}(\lambda)$ and $b_{b_{sm}}(\lambda)$, respectively]. Fig. 2.2 (from Kondratyev *et al.*, 1999) is an intercomparison of absorption cross-section spectra obtained for *non-phytoplankton* (*non-chlorophyllous*) *matter* by several workers. The absorption cross-section spectra for Lake Ontario (Canada) and Lake Ladoga (north-western Russia) (for references see Bukata *et al.*, 1995) and Chilko Lake (British Columbia) (Gallie and Murtha, 1992), and the Irish Sea (Bowers *et al.*, 1996) were determined for concentrations of *suspended minerals* in these inland water bodies, and one marginal sea, and as such are directly comparable.

The suspended inorganic matter displays a distinct increase in absorption towards short wavelengths, a less pronounced increase in the red interval, and a minimum in the range from 590 to 630 nm. Of all these determinations of $a_{sm}^{*}(\lambda)$, only the results from Lake Ladoga are numerous and thus statistically reliable because they were obtained from measurements during mid-summer in three consecutive years, and a nearly perfect coincidence between all three years of measurements has been found.

The $a_{sm}^{*}(\lambda)$ values displayed in Fig. 2.2 are in general compatible with the spectra reported by Morel and Prieur (1977) and Prieur and Sathyendranath (1981). However, in these papers the absorption cross-section spectra are either for 'suspended and dissolved material apart from algae' or for 'non-chlorophyllous particles', and hence include effects of matter other than suspended minerals. The

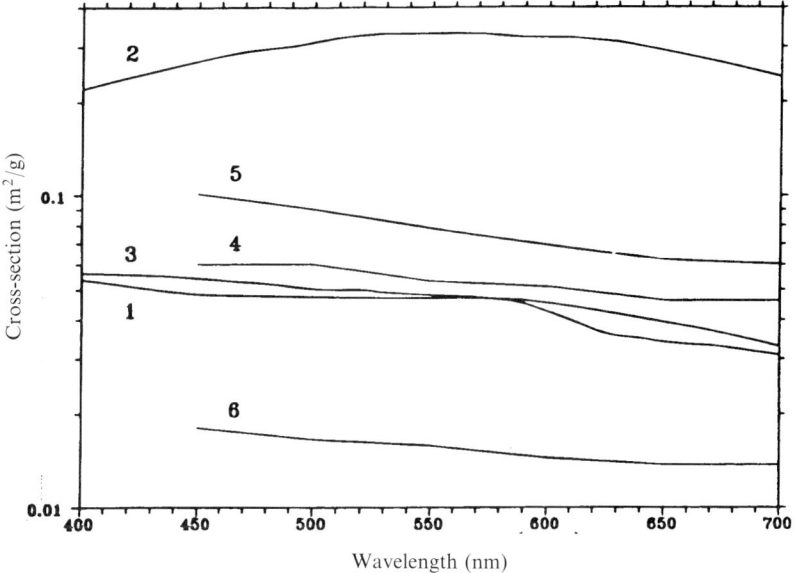

Figure 2.3. Backscattering cross-section spectra for suspended minerals. Data shown are from Bukata *et al.*, 1991 (curves 1 and 2); Gallie and Murtha, 1992 (curve 3), and Whitlock *et al.*, 1981 (curves 4, 5, and 6).

data collected by Roesler and Perry (1995) seem to come close to $a_{sm}^*(\lambda)$. Roesler and Perry also report that $a_{sm}^*(\lambda)$ decreases steadily with increasing λ, but at $\lambda \geq 600$ nm slightly begins to increase; it also increases from oceanic to coastal and fresh waters.

In their model for Case II waters Sathyendranath *et al.* (1989), suggested that the backscattering coefficient of non-chlorophyllous particles follows a λ^{-n} law with small values of n: for Case II waters, $n = 0$. Fig. 2.3 illustrates an intercomparison of the limited number of existing backscattering cross-section spectra *specifically* for suspended minerals, once again from Lake Ontario, Lake Ladoga, Chilko Lake, and three rivers in Virginia. As seen, the $b_{b_{sm}}^*$ spectra are nearly wavelength-independent/non-selective: $b_{b_{sm}}^*$ decreases only slowly with λ, which compares well with the data reported by Sathyendranath *et al.* (1989). The slight dependence of $b_{b_{sm}}^*$ on λ has been parameterized by Whitte *et al.* (1982) by:

$$b_{b_{sm}}^* = A - B \cdot (10^{-5})\lambda, \qquad (2.2)$$

with $A = 0.0316$ and $B = 0.844$, respectively. However, these coefficients are not readily transferable to a wide variety of water bodies, and should be rather considered as area-specific.

The value of $b_{b_{sm}}^*$ over the entire visible spectrum (Fig. 2.3 taken from Bukata *et al.*, 1995) varies strongly from 0.015 to 0.3 m^2/g with the highest values of $b_{b_{sm}}^*$ stemming from the waters of Lake Ladoga. For the other inland waters, $b_{b_{sm}}^*(\lambda)$ is confined to the interval 0.040–10 m^2/g, except for curve 6. It could be argued that the decrease of $b_{b_{sm}}^*$ for $\lambda \leq 490$ nm in the case of Lake Ladoga waters is a consequence of fairly high specific absorption at these wavelengths, as compared to other curves in Fig. 2.3. The enhanced absorption of *sm* in Lake Ladoga may be due to one of the following reasons:

(1) Lake Ladoga *sm* particulates are coated with colloidal *doc* (which is highly probable given the very high concentrations of *doc* in this water body ($8\,\mathrm{g\,C/m}^3 \leq C_{doc} \leq 13\,\mathrm{g\,C/m}^3$), and a high fraction of *doc* residing in the water column is in a colloidal form (see Section 2.1)). 'Coating' of *sm* particles with organic colloidal matter in non-Case I waters is amply evidenced through extensive studies conducted by Aponasenko *et al.* (1997) in a variety of natural waters.

(2) Since *doc* is known to absorb more strongly as λ decreases (see eq. (2.1)), it might be one of the reasons of high values of $a_{sm}^*(\lambda)$, and the consequent decrease of $b_{b_{sm}}^*$ at short visible wavelengths. Within spectral regions of strong absorption, there is a decrease in specific scattering: indeed, this effect has been forecast theoretically in many studies (e.g. Bricaud *et al.*, 1983; Sathyendranath *et al.*, 1989).

(3) the *sm* particulate size distribution has at least two modes, each mode having its specific mineralogical composition and, hence, its own complex index of refraction. Ulloa *et al.* (1994) examined the backscattering ratio for a collection of suspended particles in sea water for a power-law (or Junge-type) function slightly modified by K. Shifrin (1988). The backscattering ratio increases significantly with a decrease of the particle diameter (D), so that particles with $D \geq 100\,\mu\mathrm{m}$ contribute generally less than 1% to the backscattering ratio. Furthermore, the backscattering ratio (i) increases with the real part n of the refractive index m of large particles, and (ii) increases (for a given n) with the imaginary part n' of m. It was also found that the differences in the backscattering ratio in natural waters can arise from changes in the shape of the size distribution of the total particulate suspension: waters with a more negative slope in a log-log plot will have higher backscattering ratios and vice versa. In view of the *sm* data (mineralogy, characteristic sizes and complex index of refraction) for Lake Ladoga waters the combination of the above factors may cause the characteristic shape of the $b_{b_{sm}}^*$ spectrum in Lake Ladoga waters.

Based both on the above considerations and on our experimental data (see, for example, Pozdnyakov *et al.*, 1999; Bukata *et al.*, 2001) we advocate that the absorption and backscattering cross-sections established for Lake Ladoga waters are transferable only to mesotrophic and eutrophic lakes with a considerable content of *doc*.

2.2.3   Phytoplankton

Since chlorophyll pigments are among the principal natural water colorants, numerous workers have undertaken considerable efforts to obtain the optical cross-sections of chlorophyll-bearing biota. These research results revealed rather broad ranges of the numerical values of such cross-sections, and also demonstrated that the optical cross-sections are species-dependent and therefore temporally and spatially variable. In addition, the optical properties of chlorophyll-bearing biota are subject to effects of temperature, nitrogen, and light limitations. For instance, Stramski *et al.* (2002) have recently found, using the marine diatom *Thalassiosira pseudonana* as an example, that its specific absorption increases under light limitation conditions, which is due to subsequent increase in *chlorophyll*-a content in the cells. At the same time, the diatom specific absorption decreases owing to a decease in *chlorophyll*-a content as a result of temperature and nitrogen limitation. However, the intercomparison given by Bukata *et al.* (1995) indicates (their Fig. 5.1) that for very rare exceptions (e.g. the Sargasso Sea) the absorption cross-sections of phytoplankton $a^*_{php}(\lambda)$ collected by different workers from the North Central Pacific, the California coast, the upwelling region off the coast of Peru, Ladoga Lake in northern Europe, and Lake Ontario in central North America display similar spectral shapes. The data on $a^*_{php}(\lambda)$ collected by Bukata *et al.* (1995) is further substantiated, in terms of both spectral signature and magnitude, by the spectra of $a^*_{php}(\lambda)$ for oceanic, marine, and coastal waters (Doerffer, 1992; Garver and Siegel, 1997).

Surprisingly, the spectra of $a^*_{php}(\lambda)$ for Lake Ladoga and Lake Ontario are strikingly quantitatively close to each other (nearly identical), which indicates that both water bodies are characterized by optically and thus taxonomically comparable populations of chlorophyll-bearing biota.

Despite the limited data for inland waters, it should be noted that the two curves representing inland waters (curves 1 and 2 in Fig. 5.1 of Bukata *et al.*, 1995) display larger absorption cross-section values at longer wavelengths than for oceanic and coastal waters. This is fully consistent with the data reported by Roesler and Perry (1995) for oceanic, fjord, coastal, and estuarine waters. This feature is also observable in data from Chilko Lake (Gallie and Murtha, 1992), as well as from Gege (1998) and Gege *et al.* (2001). The latter workers investigated the spectral distribution of $a^*_{php}(\lambda)$ for a wide variety of algae taxa, such as cryptophyta, diatoms, dinoflagellates, and green algae indigenous to Lake Constance (Bodensee). In all these cases, the intensity of the 'red' absorption maximum was nearly comparable (but still inferior) to the height of 'blue' absorption band.

The position of the absorption bands as well as the spectral values of the specific absorption coefficients of photosynthetic pigments, and consequently the spectral values of the specific absorption coefficient of phytoplankton, reveals a spatial variability driven by a number of reasons. In particular, the optical properties of algal cells are controlled by the water body's trophic status since it largely determines, among other things, taxonomic composition and size distribution of algal cells in the phytoplankton community. Ciolli *et al.* (2002) have recently found that, in general, when phytoplankton abundance increases, larger size-classes are added incrementally to a background of smaller cells. The spectral value of $a^*_{php}(\lambda)$ proves to be very responsive to cell sizes, mostly due to pigment packaging and concentration of accessory pigments. The smaller algae species, the higher peak values of $a^*_{php}(\lambda = 450\,\text{m})$. For instance, $a^*_{php}(\lambda = 450\,\text{nm})$ is about $0.06\,\text{m}^2/\text{mg}\,\text{chl}$ for picoplankton (cells $< 2\,\mu\text{m}$), and it is *c.* 0.04, 0.03, and $0.01\,\text{m}^2/\text{mg}\,\text{chl}$ for ultraplankton (cells in the range 2–$5\,\mu\text{m}$), nanoplankton (cells in the range 5–$20\,\mu\text{m}$), and microplankton (cells $> 20\,\mu\text{m}$), respectively.

Earlier, according to the extensive studies conducted in various parts of the global ocean during various vegetation periods (April–October), at different depths (0–180 m) and in water masses of different trophic status ($0.02 < C_{chl} < 25\,\text{mg/m}^3$) (Bricaud *et al.*, 1995; Allali *et al.*, 1997), the specific absorption coefficient of phytoplankton decreases with the increasing chlorophyll concentration at all wavelengths. Based on the analysis of more than 800 spectra of $a_{php}^*(\lambda)$, the following formulation has been suggested:

$$a_{php}^*(\lambda) = A(\lambda)C_{chl}^{-B(\lambda)}, \tag{2.3}$$

where A and B are wavelength-dependent coefficients. $a_{php}^*(\lambda)$ depends on C_{chl} most strongly in the blue and red part of the spectrum. An explanation of the $a_{php}^*(\lambda)$ decline with increasing C_{chl} should be obviously sought in the interdependence between C_{chl} and the co-existing concentrations of accessory pigments: there is an experimental evidence that the content in algal cells of such pigments as non-photosynthesizing carotenoids, *chl-b* and divinyl *chl-b* is almost always higher in eutrophic rather than in mesotrophic and oligotrophic waters (Bidigare *et al.*, 1990).

In addition to variations in $a_{php}^*(\lambda)$ driven by the trophic status of a water body, seasonal variations in $a_{php}^*(\lambda)$ prove to be also very essential (Petrova, 1990; Kondratyev *et al.*, 1999; Sathyendranath *et al.*, 1999). This implies that, strictly speaking, the $a_{php}^*(\lambda)$ spectral values to be included into the hydro-optical model of a water body should be not only trophic-specific but also season-specific.

Recently Stramski *et al.* (2001) analysed the importance of taking into account the detailed composition of the planktonic community for a correct portraying of the IOPs of a targeted water body. Eighteen planktonic components covering a size range from submicrometre viruses and heterotrophic bacteria to microplanktonic species with a cell diameter of 30 μm were involved. They have shown that variations in the composition of a planktonic community might lead to significant variations in IOPs, although the chlorophyll concentration can remain unaltered.

Apart from the above factors impacting the spectral variations in $a_{php}^*(\lambda)$, the presence of the chlorophyll decomposition products in phytoplankton cells caused by oxygen hydrolysis also influences $a_{php}^*(\lambda)$ strongly. The degradation of *chl-a* into *pheophytin* results in a displacement of the absorption maximum from 430 nm to *c.* 415 nm, whereas the long-wavelength maximum at 664 nm transforms into a weaker band centred at *c.* 667 nm. A weak acidification of the medium brings about a complete disappearance of the red absorption band. Therefore, any increase in the ratio of the blue to red absorption maximum in the spectral distribution of $a_{php}^*(\lambda)$ may indicate the presence of 'dead' or dying phytoplankton in considerable amounts (Sosik and Mitchell, 1995).

In the backscattering cross-section spectra ($b_{b_{php}}^*(\lambda)$) for phytoplankton obtained for Lake Ladoga and Lake Ontario (Bukata *et. al.*, 1995) two features are distinctly discernible (their Fig. 5.3): a minimum at about 450 nm, which is followed by a broad maximum at about 550 nm. The minimum is a consequence of the scattering depression at the wavelengths of maximum absorption (430 to 450 nm): the effect that we have already discussed in relation to the *sm* backscattering depression at short wavelengths found in waters of Lake Ladoga. A similar behaviour of phytoplankton backscattering can be seen in data tabulated by Gregg *et al.* (1993) and based on modelling by Gordon for oceanic waters (see Fig. 15 in Morel and Maritorena, 2001).

The available simultaneous data (see also Roesler and Perry, 1995) on the phytoplankton absorption and backscattering coefficients, albeit very scarce, point to the great natural variability in $a_{php}^*(\lambda)$ and $b_{b_{php}}^*(\lambda)$. As an example, see Siegel *et al.* (1999) whose $a_{php}^*(\lambda)$ values for the Pomeranian Bight and the Baltic Sea are between two and ten times smaller than the $a_{php}^*(\lambda)$ values illustrated in Fig. 5.1 of Bukata *et al.* (1995). This might be a consequence of the availability of nutrients and light, the vegetation season, taxonomic composition, phytoplankton size distribution, as affected in addition by many other factors (including nutrition and light availability, zooplankton grazing, bacterial decomposition rates, etc.), weather conditions, and vertical distribution as affected also by turbulent mixing (Kondratyev *et al.*, 1999; Carder *et al.*, 1999).

Within the above discussion (Sections 2.2.2, 2.2.3), it important to add that recently Babin and Stramski (2002) obtained experimental evidence indicating that phytoplankton cells as well as non-living organic particles, mineral particles and mixtures of organic and inorganic particles, all exhibit similar absorbance pattern in the *near-infrared* spectral region above 700 nm, that is, a nearly flat spectrum with values very close to zero. This finding appears very significant for the analysis of formation of spectral distribution of water-living radiance at these wavelengths, and hence, for the solution of the inverse problem in remote sensing of natural waters, especially for achieving accurate atmospheric correction of satellite images in the visible (see Chapter 5).

2.2.4 Dissolved organic matter (*doc*)

Although the exponential in eq. (2.1) is also variable in the world oceans, there is better agreement in spectral shape among the measured cross-sections of *doc* (for references see Bukata *et al.*, 1995; Kondratyev *et al.*, 1999) compared to *chl* and *sm* absorption cross-sections.

For Lake Ontario the absorption cross-section for *doc* follows (Bukata *et al.* 1995):

$$a_{doc}^*(\lambda) = 0.173 \exp[-0.0157(\lambda - 400 \text{ nm})]. \qquad (2.4)$$

Also following an exponential function, the spectral distribution of $a_{doc}^*(\lambda)$ for Lake Ladoga (Kondratyev *et al.*, 1990) is characterized by higher values over the spectral region studied. This is thought to be due to the aforementioned colloidal fraction of *doc* in Lake Ladoga.

2.2.5 Synthesis

Even a cursory analysis of the above data on absorption and backscattering cross-sections of the major CPAs, debatably chosen by us in the preceding section as the main components of a hydro-optical model of Case II waters, indicates that the model should be statistically substantiated and result from a weighted summation of the optical properties of the ensemble of phytoplankton species, suspended mineral particulates and dissolved organic substances co-existing in non-Case I waters.

We also argue here that really representative values of absorption and back-scattering cross-sections for the CPAs are difficult to obtain in the laboratory. Indeed, sampling followed by numerous technological procedures is always like a vivisection of the natural aquatic medium, and hence, cannot account for the actual *status quo* in the aquatic medium. Besides, undersampling, especially in optically highly heterogeneous water bodies (and Case II waters in this sense are frequently highly variable in space) is generally a limitation in field campaigns, which does not allow reliable statistics. The other argument in support of this point of view is the fluorescence produced by *living* phytoplankton and *doc*. As will be shown in Chapters 3 and 4, fluorescence by phytoplankton and *doc* is capable of essentially changing the optical properties of these CPAs. Determination of the IOPs of CPAs *in vivo/in situ* offers in this regard an incontestable advantage.

Based on a multivariate optimization technique (see Chapter 3), the retrieval of absorption and backscattering cross-sections of all three CPAs from *in situ* determinations of their concentrations and simultaneous measurements of the volume reflectance coefficient $R(\lambda, -0)$, is probably a better approach. Indeed, being a convolution of optical contributions of all CPAs existing as a composition of assemblages of each type of CPA, $R(\lambda, -0)$ will eventually provide through a multivariate optimization technique a determination of *resultant* cross-sections accounting for a weighted optical action of each member of the CPA assemblages.

This 'R' technique is less time-consuming, is less susceptible to instrumental errors/artefacts, and can be applied to assure higher spatial resolution generally unattainable for the method of laboratory determinations of CPA cross-sections.

The Lake Ladoga hydro-optical model (Fig. 2.4 and Table 2.2) has been obtained with the above 'R' method. Ship cruises covered the entire coastal region of the lake (which is the most optically heterogeneous part) and vast pelagic provinces. The field campaign has been conducted in the mid-summer vegetation season for three consecutive years following one and the same scheme. The spectra of absorption and backscattering by the CPAs were very close each year.

As the cross-sections presented in Table 2.2 and in Fig. 2.4 are a result of a 3-year averaging, they give a sufficient statistical credit to the developed hydro-optical model. Moreover, we showed above that the Lake Ladoga spectral distributions and magnitudes of absorption and backscattering cross-sections are generally similar to those obtained for other water bodies, for instance, Lake Ontario, which from many perspectives is similar/affinitive to Lake Ladoga (Kondratyev *et al.*, 1999). These considerations led us to adopt the Lake Ladoga hydro-optical model as a basis for the numerical experiments to be discussed in the following chapters.

It should be underscored that the 'R' method is certainly not free from drawbacks. One of those is due to the fact that we attribute the formation of the measured 'R' spectrum to a *limited* number of players/water constituents, which in

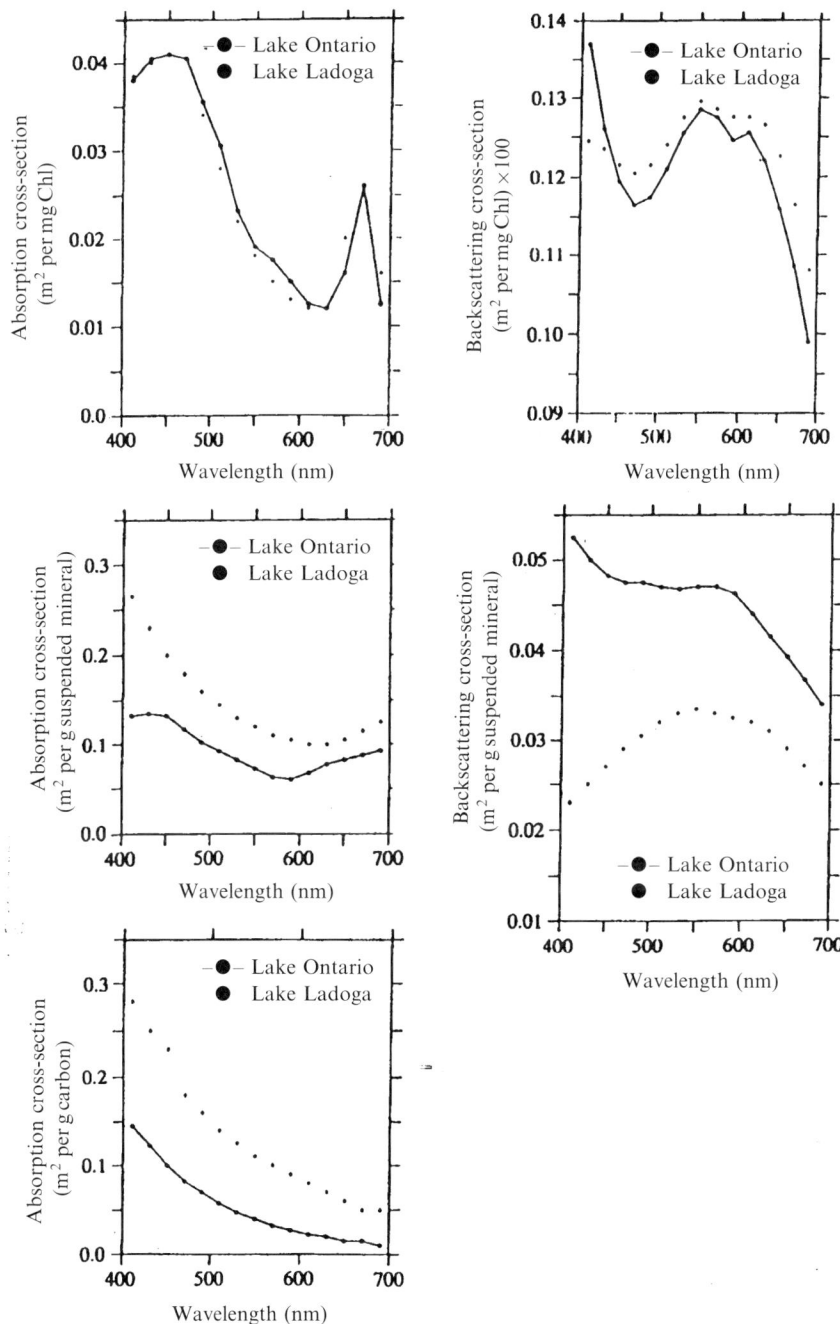

Figure 2.4. Specific spectral absorption and backscattering coefficients (cross-sections) for phytoplankton (*chl*), *sm*, and *doc* determined for Lake Ladoga (●) and Lake Ontario (–●–) waters (Bukata *et al.*, 1995).

Table 2.2. Spectral absorption and backscattering cross-sections for phytoplankton, suspended minerals, and dissolved organics in Lake Ladoga waters.

λ (nm)	a^*_{php} (m^2/mg)	a^*_{sm} (m^2/g)	a^*_{doc} (m^2/gC)	$b^*_{b_{php}}$ (m^2/mg)	$b^*_{b_{sm}}$ (m^2/g)
410	0.038 00	0.265 00	0.280 00	0.012 40	0.23 00
430	0.040 00	0.230 00	0.250 00	0.012 30	0.25 00
450	0.041 00	0.200 00	0.230 00	0.012 10	0.27 00
470	0.040 00	0.180 00	0.180 00	0.012 00	0.29 00
490	0.034 00	0.160 00	0.160 00	0.012 10	0.30 50
510	0.028 00	0.145 00	0.140 00	0.012 40	0.32 00
530	0.022 00	0.130 00	0.125 00	0.012 70	0.33 00
550	0.018 00	0.120 00	0.110 00	0.012 90	0.33 50
570	0.015 00	0.110 00	0.100 00	0.012 80	0.33 00
590	0.013 00	0.105 00	0.090 00	0.012 70	0.32 50
610	0.012 00	0.100 00	0.080 00	0.012 70	0.32 00
630	0.012 00	0.100 00	0.070 00	0.012 60	0.31 00
650	0.020 00	0.105 00	0.060 00	0.012 20	0.29 00
670	0.025 00	0.115 00	0.050 00	0.011 60	0.27 00
690	0.016 00	0.125 00	0.050 00	0.010 80	0.25 00

reality are more numerous. It implies that the retrieved absorption and backscattering cross-sections, being attributed to a selected number of CPAs, should accommodate in their spectral signatures some elements arising from 'unaccounted' CPAs. However, as seen from above, the typical spectral features of absorption and backscattering cross-sections of phytoplankton, *doc*, and *sm*, nearly strictly agree with the optical properties of these CPAs, known from other waters, and thus explicitly indicate that the contributions from other ('unaccounted') CPAs should be small.

This is one of the reasons why we confined our hydro-optical model to only four components. Although there are also data available on cross-sections of bacterioplankton, detritus, and air bubbles (Morel and Ahn, 1991; Stramski and Kiefer, 1991; Marks *et al.*, 2001; for other references, see also Kondratyev *et al.*, 1999) which could be included into a more exhaustive hydro-optical model, we abstained from it for the following reasons. Firstly, the data on the above additional constituents/CPAs are scarce and, hence, hardly reliable. Secondly, inclusion of an extended number of CPAs does not necessarily lead to more accurate retrieval results: the potentials of multivariate optimization techniques (as well as many others alike) decline with the increasing number of unknowns (Kondratyev *et al.*, 1990). It means that some more sophisticated retrieval approaches (see, for example, Maritorena *et al.*, 2002) are required, which are, however, bad candidates for operational use when processing large-scale space images.

3

Colour formation in natural waters: numerical simulations of the forward problem

Beginning with a very successful interpretation of spectral images from early LANDSAT satellites in the 1970s, and particularly from the CZCS since 1978, water colour became an important remotely sensed parameter. This colour information can later be used for the assessment of water quality characteristics, the trophic status of natural water bodies, the primary production rate, etc. (Multi-author, 2000). The satellite sensors operating at present, e.g. SeaWiFS, MODIS, MOS, MERIS, and others coming soon, aim to study the marine biosphere via water colour to achieve some far-reaching goals including the assessment of global oceanic/marine/fresh water primary production, as well as carbon uptake or emission (NASA, 1993).

In oceanology, the principal water colour characteristics are frequently documented on a routine basis in order to obtain operational data on water purity (Jerlov, 1976). At the same time, in classical limnology, the water colour, as assessed through employing special colour scales (by no means based on physical approaches), is also considered to be a very informative index, closely related to both hydrochemical and eutrophication processes (Hutchinson, 1957). To the present time, large amounts of relevant data have been and are still being collected for ocean basins, lakes, reservoirs, and large rivers throughout the world.

This wide use of water colour in remote sensing and in oceanology and limnology as a parameter characterizing the state of the world oceans and inland water bodies merits a close look at its sensitivity to variations in water quality, including its responsiveness to variations in the concentration of such major CPAs as phytoplankton (*chl*), suspended minerals (*sm*) and dissolved organic carbon (*doc*), and bottom albedo (in optically shallow waters). Furthermore, it appeared important to us to investigate through numerical simulations the appropriateness of using water colour as a parameter, which unequivocally indicates, at least, the predominance of one or another CPA.

In Chapter 1, we pointed out that the resulting radiometric water colour is a convolution of all photon interactions with the aquatic medium. These interactions encompass absorption, elastic and inelastic (Raman) scattering of solar radiation, fluorescence by *chl* and *doc*, and reflection from the bottom (in optically shallow waters).

In what follows, we pursue a step-by-step approach, which can also be seen as 'progression from the simple to the more complex'. Therefore, we will start with an optically semi-infinite aquatic medium considering only absorption and elastic scattering. This then will be followed by the inclusion of trans-spectral processes. Finally, bottom albedo will be involved, in order to analyse the impact of different types of bottom cover on the spectrum of upwelling radiance (eventually captured by a satellite sensor), and hence on the radiometric water colour.

3.1   OPTICALLY SEMI-INFINITE AQUATIC MEDIUM

3.1.1   Elastic scattering

Following Austin (1974), the upwelling radiance just above the water–air interface at any wavelength, $L_u(+0)$, is related to the inherent optical properties of the water column through the volume reflectance just beneath the air–water interface, $R(-0)$, via

$$L_u(+0) = R(-0)\lfloor 1 - \rho_\uparrow(\theta)\rfloor E_d(+0)(1 - \rho_\downarrow)/Qn^2[1 - 0.48R(-0)], \qquad (3.1)$$

where $\rho_\uparrow(\theta)$ = internal reflectivity for the in-water-refracted angle θ corresponding to the remote sensing viewing angle; $\rho_\downarrow$ = surface reflectivity for downwelling irradiance, $E_d(+0)$; n = relative index of refraction of water to air; $E_u(-0)$ = upwelling irradiance just beneath the water–air interface; $Q = E_u(-0)/L_u(-0)$ with $E_d(-0)$ = downwelling irradiance and $L_u(-0)$ = upwelling radiance, both being just beneath the air–water interface.

Given the spectrum of $E_d(+0)$, the variations in $L_u(+0)$ due to variations in the concentrations of CPAs can be accounted via the spectral variations of the volume reflectance, $R(-0)$, which, in turn, is a function of the composition of water constituents (see Chapter 2).

The four-component bio-optical model suggested by us (see Table 2.1 (columns $a_w^{PF}(\lambda)$, and b_w^{MA}) and Table 2.2) that considers pure water (w), phytoplankton (chl), *doc*, and *sm* as the principal aquatic CPAs, relates the bulk absorption and back-scattering coefficients $a(\lambda)$ and $b_b(\lambda)$, to the concentrations of these CPAs and their specific absorption and backscattering coefficients (optical cross-section spectra) $a^*(\lambda)$ and $b_b^*(\lambda)$:

$$a(\lambda) = a_w(\lambda) + C_{chl}a_{chl}^*(\lambda) + C_{sm}a_{sm}^*(\lambda) + C_{doc}a_{doc}^*, \qquad (3.2)$$

$$b_b(\lambda) = b_{b_w}(\lambda) + C_{chl}b_{b_{chl}}^*(\lambda) + C_{sm}b_{b_{sm}}^*, \qquad (3.3)$$

where C_{chl}, C_{sm}, and C_{doc} = concentrations of *chl*, *sm*, and *doc*, respectively.

Table 3.1. Variations in C_{chl} (µg/l), C_{sm} (mg/l), and C_{doc} (mgC/l) used in the present numerical simulation studies of the λ_{dom} dependence on the concentration vector C.

Discrete sequence of			Quasi-continuous sequence of		
C_{chl}	C_{sm}	C_{doc}	C_{chl} ($\Delta C_{chl} = 0.5$)	C_{sm} ($\Delta C_{sm} = 1.0$)	C_{doc} ($\Delta C_{doc} = 0.5$)
0; 1; 2;		0; 1; 5;		0–20	
5; 10; 20		10			
0; 1; 5;	0; 0.1;				0–10
10; 20	0.2; 0.5;				
	1; 2; 5;				
	10				
0; 0.2;	0; 0.1;				0–10
0.5; 1;	0.5; 1.0;				
2; 5; 10;	5; 10;				
20					
	0; 0.1;	0; 0.1	0–20		
	0.2; 0.5;	0.5; 1			
	1; 2; 5;	5; 10			
	10				

Developed by Jerome *et al.* (1988) and being applicable to a wide range of organic and inorganic CPAs of inland and coastal waters, eqs. (1.18) and (1.19) were employed to relate, at any wavelength, the apparent optical property volume reflectance $R(-0)$ to the IOPs of the water column, namely the bulk absorption coefficient a and the bulk backscattering coefficient b_b.

Using eqs (1.1–1.6), the CIE colour mixture values for $x(\lambda), y(\lambda)$ and $z(\lambda)$, and the spectrum $E_d(\lambda, +0)$ inferred from a standard atmospheric radiative transfer model such as discussed by Gregg and Carder (1990), we have conducted numerical simulations of the dominant wavelength λ_{dom} for the water composition characteristic of Lake Ladoga (Table 3.1). Only the nadir view was considered and the sun zenith angle θ_0 was set to $30°$. The water surface was assumed flat and the CPAs were assumed vertically homogeneous. The value of Q was set to π, which, of course, is a rough approximation (see Chapter 1), but given that we consider here exclusively the nadir view, this assumption should not have major consequences. The dependence of the dominant water colour wavelength λ_{dom} on the varying concentration vector C, i.e. the set of C_{chl}, C_{sm}, and C_{doc} (see Table 3.1), is presented as 3-D plots in Figs 3.1–3.4 (Pozdnyakov *et al.*, 1998). For pure water (i.e. C_{chl}, C_{sm}, and $C_{doc} = 0$), λ_{dom} is found at 472 nm. For the Lake Ontario model (see Fig. 2.4) $\lambda_{dom} \approx 471$ nm. Jerlov (1976) reports that in the clearest ocean waters $\lambda_{dom} \approx 465$–475 nm. Although the calculated value of λ_{dom} is within this spectral interval, the existence of natural waters displaying λ_{dom} as low as 465 nm requires a

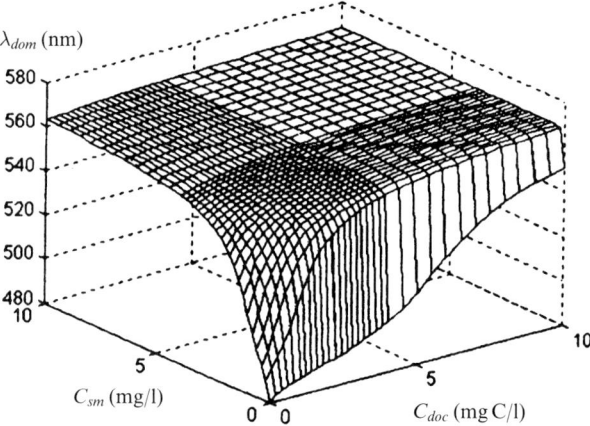

Figure 3.1. Dependence of λ_{dom} (nm) on variations in C_{sm} (mg/l) and C_{doc} (mg C/l) at $C_{chl} = 1\,\mu g/l$.

closer look. One of the obvious reasons resides in the different spectral composition of $E_d(\lambda, +0)$ for different latitudes.

The analyses of Figs 3.1–3.4 indicate that λ_{dom} *asymptotically* tends to a certain limit-value $\lambda_{dom,lim}$ (or *end-point*) with increasing concentrations of certain PCAs. This *end-point* for simultaneously abnormally high concentrations of *chl*, *sm*, and *doc* (Table 3.2) is at about 571 nm. It reaches an absolute maximum very close to 577 nm at equally abnormally high C_{chl} and C_{sm} (≥ 1000, in respective units). The use of the cross-sections from Lake Ontario together with the values of $E_d(\lambda, +0)$ as inferred from a standard atmosphere optical model yields slightly different values of $\lambda_{dom,lim}$ (Table 3.2). Nevertheless, both numerical modelling results unanimously point to this fundamental property of the colour of natural water bodies to approach an end-point. This finding was recently substantiated by Doxaran *et al.* (2002) who showed that the water volume reflectance in the first two SPOT wavebands (500–590 nm and 610–680 nm) tends to an end-point as the concentration of suspended particulate matter in water exceeds a threshold of about 100 mg/l.

Fig. 3.5 illustrates the loci of all calculated water colours within the chromaticity diagram envelope for the optical cross-section spectra and CPA concentrations (Table 3.1) typical of Lake Ladoga. λ_{dom} ranges from 471 nm for chemically pure water devoid of any organic or inorganic matter to ~ 575 nm ('end-point', asymptotically approached as CPA concentrations were increased either individually or collectively). The experimentally determined loci of the (X, Y) chromaticity coordinate pairs appropriate to the Lake Ladoga water and generated from the optical cross-section spectra appropriate to Lake Ladoga CPAs are consistent with the modelled results of Fig. 3.5 within a restricted dominant wavelength range $563 \leq \lambda_{dom} \leq 570$ nm.

In a similar manner, using the optical cross-section spectra determined for Lake Ontario (results are not shown), the numerical model, regarding its 'end-point' colour and colour purity predictions, displayed remarkably good agreement with

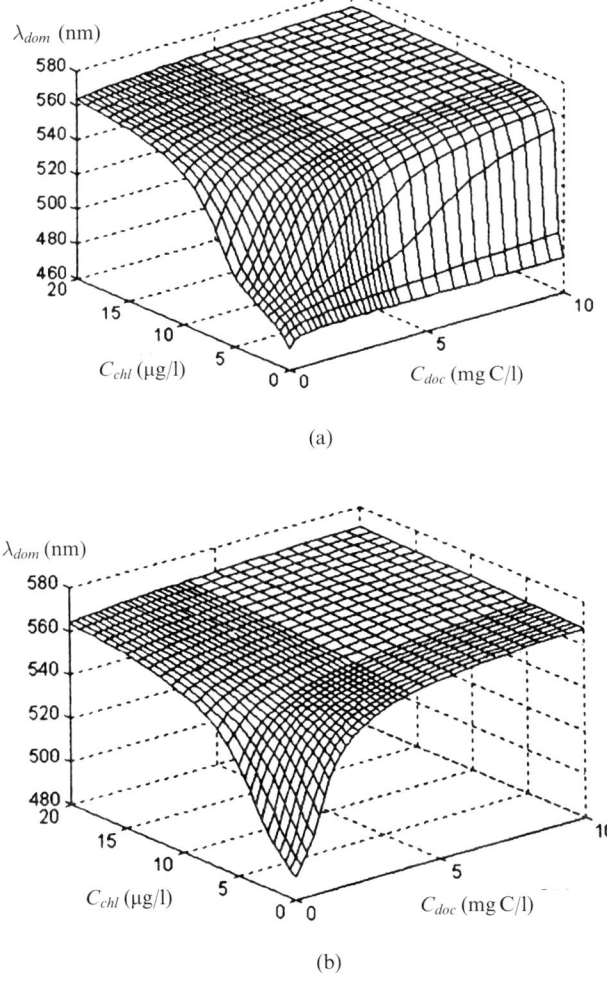

Figure 3.2. Dependence of λ_{dom} (nm) on variations in C_{chl} (µg/l) and C_{doc} (mg C/l) at $C_{sm} = 0$ (a) and 0.4 mg/l (b).

direct observations of water colour and quality in the diverse geographic regions of Lake Ladoga (Russia), Lake Ontario (central Canada), and British Columbia (western Canada) (Bukata *et al.*, 2001).

In order to investigate its possible universality, additional optical and water quality data sets were considered, namely subsurface radiometric measurements of upwelling and downwelling irradiances in the visible and direct measurements of CPA concentrations in two more of the Laurentian Great Lakes (Lakes Erie and Michigan) and a number of relatively small boreal lakes in northern Ontario. Several lakes were sampled in the Sudbury, Algoma, and Dorset areas (Tanis and Marshall, 1989). Since the optical cross-section spectra for these lakes were not available, direct

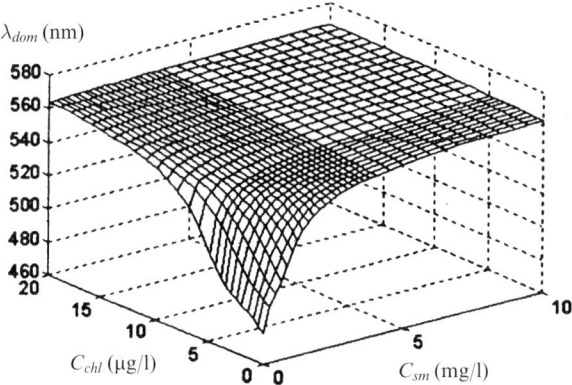

Figure 3.3. Dependence of λ_{dom} (nm) on variations in C_{chl} (µg/l) and C_{sm} (mg/l) at $C_{doc} = 0.4\,\mathrm{mgC/l}$.

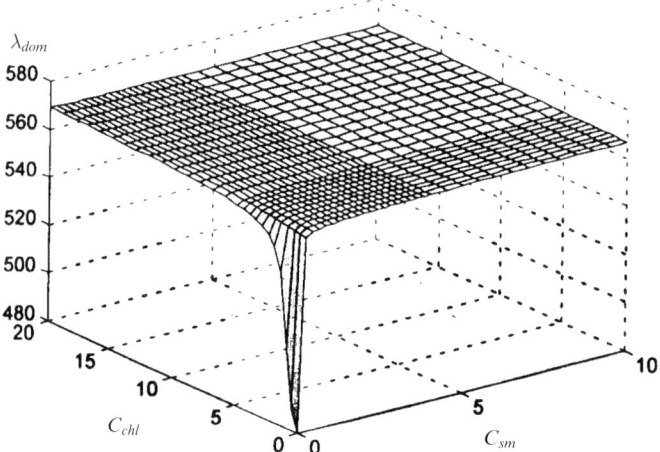

Figure 3.4. Dependence of λ_{dom} on variations in C_{chl} (µg/l) and C_{sm} (mg/l) at $C_{doc} = 5\,\mathrm{mg\,C/l}$.

measurements of global radiation and the subsurface volume reflectance were used to generate the (X, Y) chromaticity coordinate pairs for these lakes. Lakes Michigan and Erie displayed λ_{dom} values between 565 and 570 nm (comparable to those displayed by Lakes Ladoga and Ontario) as CPA concentrations increased either individually or collectively.

Table 3.3 lists the *in situ* measurements of the CPA concentrations along with the calculated chromaticity coordinates X (red) and Y (green), using the upwelling radiance spectra obtained over several northern Ontario boreal lakes and the two additional Laurentian Great Lakes. Also included in Table 3.3 are the analogous data of the European lakes Krasnoye (Russia), Zug (Switzerland), and Lucerne (Switzerland). Figure 3.6 displays the loci of all the derived (X, Y) coordinate

Table 3.2. Estimates for end-point aquatic colours of natural waters and associated spectral purities.

C_{chl} (µg/l)	C_{sm} (mg/l)	C_{doc} (mg C/l)	λ_{dom} (nm)		p (%)	
0.00	0.00	0.00	471*	472**	83*	77**
0.00	0.00	1000.00	478*	478**	10*	18**
0.00	1000.00	0.00	575*	569*	27*	33**
1000.00	0.00	0.00	582*	577**	48*	42**
1000.00	0.00	1000.00	583*	576**	35*	37**
0.00	1000.00	1000.00	576*	571**	35*	37**
1000.00	1000.00	0.00	574*	570**	30*	35**
1000.00	1000.00	1000.00	576*	572**	37*	38**

* Lake Ontario cross-sections and $E_d(+0, \lambda)$ inferred from model simulations.
** Lake Ladoga cross-sections and $E_d(-0, \lambda)$ measured *in situ*.

pairs. λ_{dom} values in a wide range from 485 nm to 574 nm are evident, in agreement with the modelled values of Fig. 3.5 and the Lake Ladoga observations.

We have used direct observations of radiometric colour and local CPA concentrations in a variety of inland waters in four countries (Canada, Russia, USA, and Switzerland) in order to illustrate the validity and the universality of a radiometric water colour model that relates the colour of a water body to the organic and inorganic suspended and dissolved colour-producing agents (CPA) indigenous to that water body. The linkages between aquatic colour and aquatic CPAs are the specific spectral absorption and scattering coefficients (optical cross-section spectra) of those indigenous CPA. The water colour model illustrates that:

(a) the ultimate 'end-point' colour of all natural water bodies can be defined by a dominant wavelength, λ_{dom} of ~ 572 nm (yellow-green region of the visible spectrum), which, when considered in concert with an 'end-point' colour purity, p, of only $\sim 40\%$, would shift the perceived end-point colour of all natural waters to brown as CPA concentrations increase either individually or collectively,

(b) chromaticity analyses, which have been successfully utilized in extracting oceanic chlorophyll concentrations from remote measurements of the colour of Case I waters, are of limited use for remote sensing of inland and coastal (non-Case I) waters.

That a 'universal' use of Lake Ontario and Lake Ladoga optical cross-section spectra would produce such surprisingly consistent results for water colour in geographically very distant waters certainly warrants further discussion, which, for brevity, will be confined here to the Laurentian Great Lakes, although an analogous one could be presented for the case of the North European water bodies. Suspended inorganic particulate matter in natural water bodies generally reflects the geologic structure of the basin confining the lake (relatively long residence time) or the river system (relatively short residence time).

Table 3.3. Summary of measured CPA concentrations and the chromaticity coefficients (X and Y) for some lakes considered in this study resulting from the use of direct measurements of global radiation and subsurface volume reflectance. The concentrations of *chl*, *sm*, and *doc* are given in μg/l, mg/l, and mg C/l, respectively.

Lake	*chl*	*doc*	*sm**	*X*	*Y*
Smoothwater	0.5	1.3	0.1	0.301	0.373
Smoothwater	0.8	1.7	0.1	0.312	0.363
Whitepine	1.5	2.4	0.1	0.351	0.401
Whitepine	0.65	2.0	0.1	0.330	0.401
Whitepine	0.5	1.6	0.1	0.310	0.392
Sunnywater	0.1	0.2	0.1	0.193	0.311
Sunnywater	0.2	0.7	0.1	0.194	0.275
Sunnywater	0.35	0.3	0.1	0.182	0.261
Sunnywater	0.35	0.3	0.1	0.179	0.265
Wolf	0.18	0.7	0.1	0.251	0.353
Wolf	0.2	0.7	0.1	0.251	0.350
Wolf	0.11	0.4	0.1	0.194	0.311
Wolf	0.3	0.7	0.1	0.190	0.302
North Yorkston	0.37	1.2	0.1	0.312	0.381
North Yorkston	0.4	1.3	0.1	0.292	0.367
Dougherty	0.14	0.7	0.1	0.270	0.353
Dougherty	0.2	0.7	0.1	0.271	0.351
Dougherty	0.2	0.4	0.1	0.282	0.302
Centre	0.65	1.8	0.1	0.335	0.401
Centre	0.6	2.0	0.1	0.332	0.395
Centre	0.2	0.4	0.1	0.212	0.302
Centre	0.82	2.0	0.1	0.332	0.381
Erie	18.1	2.5	2.0	0.371	0.410
Erie	21.1	2.5	2.3	0.386	0.415
Michigan	7.7	2.5	0.52	0.361	0.392
Michigan	6.2	2.5	0.31	0.355	0.381
Ontario	5.63	2.3	3.4	0.372	0.381
Krasnoe	2.9	1.8	0.4	0.369	0.395
Zug	4.6	2.5	1.63	0.342	0.372
Lucerne	1.5	0.93	1.4	0.330	0.362
Ladoga	7.1	8.0	0.8	0.379	0.409

*Concentrations of *sm* given as 0.1 indicate, in reality, that they are not in excess of this value.

The Laurentian Great Lakes basins, while displaying some geologic variability, also show a geologic consistency as well as connections as part of the St. Lawrence River basin. Lakes Ladoga and Ontario (Bukata *et al.*, 1995) possess chlorophyll-bearing biota whose scattering and absorption cross-section spectra are very similar (but distinctly different from those of Case I water biota), although their respective suspended particulates display scattering and absorption cross-section spectra are markedly different.

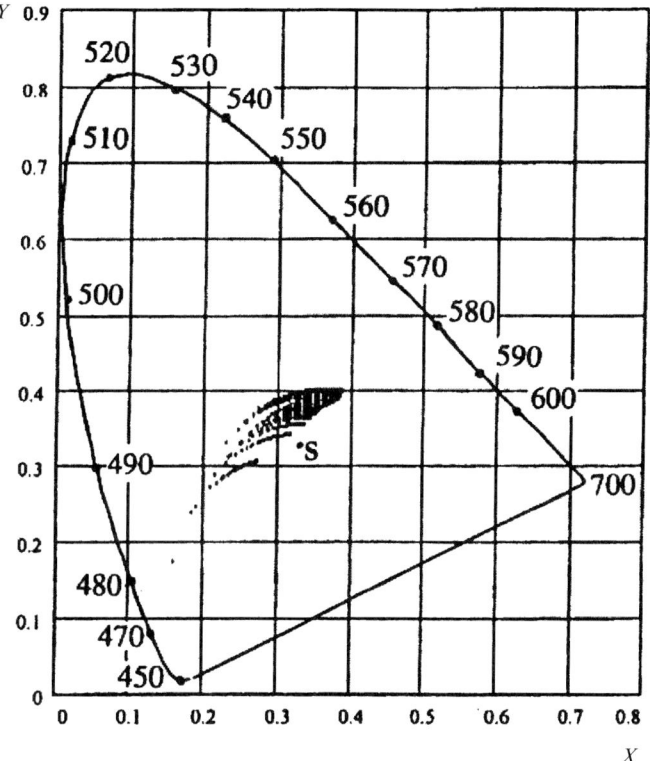

Figure 3.5. The Y (green) and X (red) chromaticity coordinates resulting from the totality of modelled radiometric colour for a wide range of hypothetical combinations of aquatic CPAs defined by the cross-section spectra appropriate to the CPAs indigenous to Lake Ladoga.

As seen in the preceding chapter, very few optical cross-section spectra exist for inorganic CPAs indigenous to natural waters. However, if compared, the Lake Ontario particulates display optical cross-section spectra that are similar (although not identical) to those obtained for Chilko Lake, British Columbia, Canada (Gallie and Murtha, 1992) and rivers in Virginia, USA (Whitlock *et al.*, 1981). Although the specific absorption coefficient for indigenous dissolved organic matter is highly variable in Canadian waters (Jerome *et al.*, 1998a), the shape of the *doc* absorption cross-section spectra is mathematically well defined by an exponential function decreasing with increasing wavelength. The boreal lakes in northern Ontario, while again displaying some variability of geological structure, have strong geologic similarities with the Great Lakes basin and northeastern USA.

The use of CPA concentrations and their optical cross-section spectra from Lake Ontario as a 'surrogate' data set, therefore, yield comparable chromaticity coordinates for all water columns whose dominant CPAs were other than *sm*, because freshwater algae in the large Lakes Ladoga and Ontario display very similar scattering and absorption cross-section spectra, and essentially all aquatic *doc* mixtures

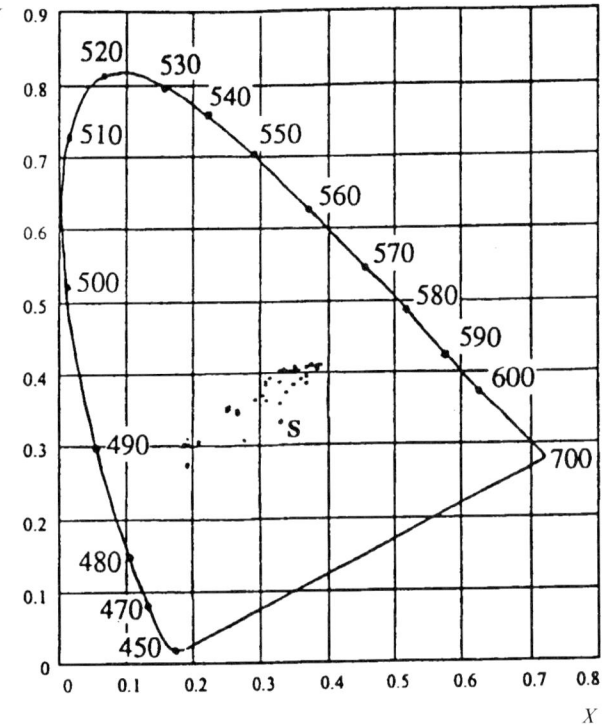

Figure 3.6. The Y (green) and X (red) chromaticity coordinates resulting from the use of direct measurements of global radiation and subsurface volume reflectance in conjunction with the optical cross-section spectra appropriate to Lake Ontario CPAs for the North American lakes and with the optical cross-section spectra appropriate to Lake Ladoga CPAs for the European lakes listed in Table 3.3.

display a comparable shape of the specific absorption spectra. Although the use of optical cross-section spectra different from those of Lake Ontario (particularly for suspended particulates) would result in slightly different 'end-point' aquatic colour and 'end-point' colour purity, it is reasonable to presume that such differences, in most cases, would not be large and that the model used here would remain intact. Most significant impacts to 'end-point' values would be observed for water bodies containing inorganic CPAs with cross-section spectra that were substantially different from those indigenous to Lake Ontario (e.g. white sand or glacial flour).

To further investigate the limits of applicability of the cross-section spectra used herein, the radiometric colour model along with the direct measurements of global radiation was then used jointly with the measured CPA concentrations from Table 3.3 and both the optical cross-sections of Lakes Ladoga and Ontario to yield the geographic regrouping in Fig. 3.6. The chromaticity coordinates for North American lakes are plotted in Fig. 3.6 using the optical cross section spectra of Lake Ontario. The chromaticity coordinates for North European lakes

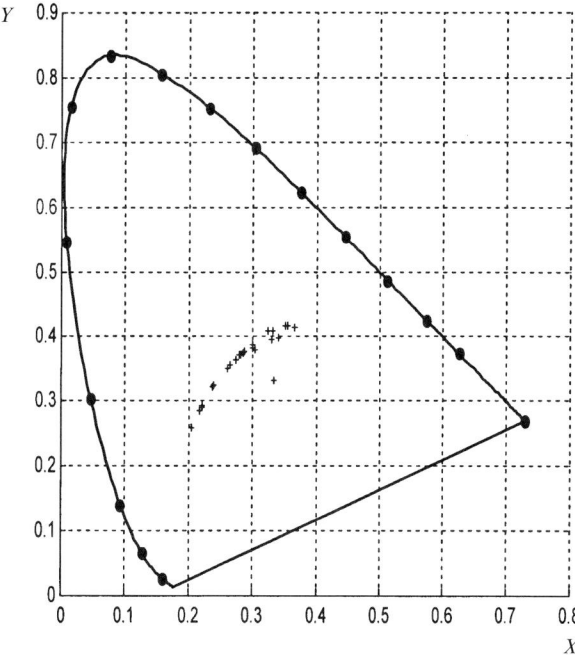

Figure 3.7. The Y (green) and X (red) chromaticity coordinates (vertical and horizontal axes, respectively) for the Great Lakes, European lakes, and boreal lakes of northern Ontario listed in Table 3.3, resulting from the use of direct measurements of global radiation and subsurface volume reflectance in conjunction with the optical cross-section spectra appropriate to CPAs of Lake Ontario.

are plotted in Fig. 3.7 using the optical cross-section spectra for Lake Ladoga. Similarities between Figs 3.6 and 3.7 are evident. This result can be cautiously considered as an indication that for at least inland water bodies, where aquatic colour is predominantly controlled by *doc* and *sm* of terrestrial origin, a judicious selection of watersheds displaying comparable geologic origin/history and latitude enables the use of *surrogate* spectral absorption and scattering cross-sections. Such surrogate cross-sections were used (Pozdnyakov *et al.*, 1999) to restore observed water leaving radiance spectra in the inner and outer parts of Saginaw Bay of Lake Huron.

   Returning to the numerical experiments discussed at the beginning of this section, it is worth noting that in addition to the existence of *end-points*, natural water colour can exhibit another important feature. The graphs in Fig. 3.8 provide clear evidence that waters devoid of *sm* and containing *chl* in extremely small amounts (below $0.2\,\mu g/l$) vary slightly in colour provided C_{doc} is less than $0.5\,mg\,C/l$. As soon as C_{doc} is over this limit, the colour no longer varies and $\lambda_{dom} = 478\,nm$, even if C_{doc} assumes abnormally high values ($\sim1000\,mg\,C/l$). However, this effect disappears completely when C_{sm} reaches $0.1\,mg/l$. Even without algae ($C_{chl} = 0$) (Fig. 3.2(b)), already a small increase in C_{sm} results in a

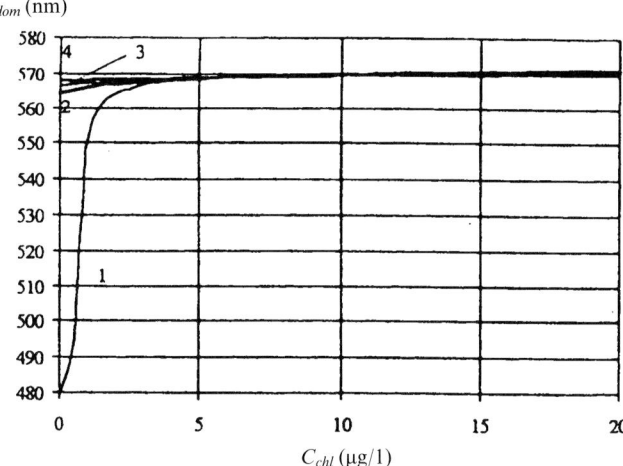

Figure 3.8. Dependence of λ_{dom} (nm) upon the variations in C_{chl} (µg/l) at $C_{doc} = 10\,\text{mg C/l}$ and for four different C_{sm} (mg/l) values: 0 (**1**), 0.1 (**2**), 0.3 (**3**), and 10 (**4**).

very substantial increase of λ_{dom}. Simultaneously, the value of λ_{dom} rapidly shifts to the *end-point*, whereby the sensitivity of λ_{dom} to C_{doc} increments becomes more and more restricted to the range of lower C_{doc} concentrations. These trends in λ_{dom} become further accentuated when water is enriched in *chl* or *sm* (Fig. 3.2(a), (b)), e.g. at $C_{sm} = 1.0\,\text{mg/l}$, the transition to the asymptotic regime occurs at C_{doc} between 4 and 5 mg C/l, whereas at $C_{sm} = 0.1\,\text{mg/l}$ it takes place at $C_{doc} \approx 8\text{–}9\,\text{mg C/l}$.

A close analysis of the shift of λ_{dom} to the end-point reveals that λ_{dom} sensitivity to variations in C_{sm} is strongly controlled by the abundance in C_{doc}: higher C_{doc} results in a rapid shrinking of the C_{sm} range within which λ_{dom} responds to increments in C_{sm}.

At the same time, there is a less rapid transition of λ_{dom} to the *end-point* for a simultaneous increase of C_{chl} and C_{sm} when C_{doc} is zero. However, with growing C_{doc}, this transition takes place at successively lower values of C_{chl}. It should also be noted that at $C_{chl} = 0$ and $C_{doc} \geq 10\,\text{mg C/l}$, there is a characteristic step-like increase of λ_{dom} from about 480 nm to 565 nm, when C_{sm} increases from zero to 0.1 mg/l (Fig. 3.8). This nearly step-like increase is less rapid with growing C_{chl}: already at $C_{chl} > 2\,\text{µg/l}$ it disappears.

The aforementioned shrinking of the possible λ_{dom} range with increasing CPA concentrations is also accompanied by a considerable narrowing of the range of $C_{chl}/C_{sm}/C_{doc}$ combinations leading to a particular λ_{dom}, as illustrated in Fig. 3.9: the range of C_{sm} and C_{chl} values resulting in $\lambda_{dom} < 561$ nm, drastically decreases with increasing C_{doc}.

Table 3.4 lists calculated $C_{chl}/C_{sm}/C_{doc}$ combinations (within the scope of our numerical modelling assumptions) leading to $470 < \lambda_{dom} < 500$ nm. Consequently, all other combinations in CPA concentrations specified in Table 3.4 would result in $\lambda_{dom} \geq 500$ nm. Natural waters with $500 < \lambda_{dom} < 575$ nm contain either

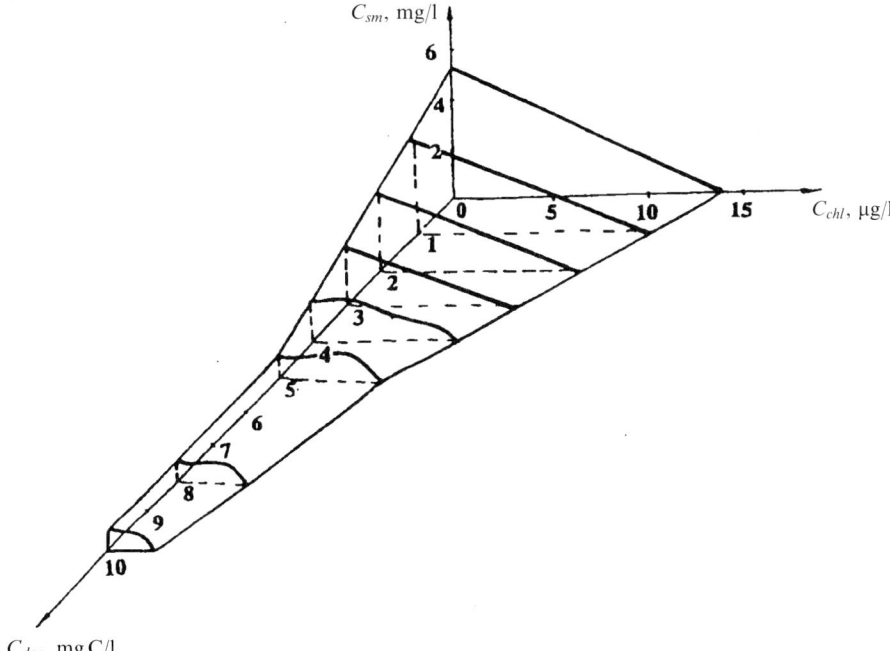

Figure 3.9. Ranges of C_{sm} and C_{chl} resulting in $\lambda_{dom} < 561$ nm at C_{doc} varying from 0 to 10 mg C/l.

$C_{chl} > 10\,\mu g/l$ or $C_{sm} > 6\,mg/l$ or $C_{doc} > 6\,mg\,C/l$ (if other CPAs are absent). The presence, even in very small amounts, of any other optically active component (i.e. additional to *chl* or *sm* or *doc*) reduces substantially the concentration limits identified above.

This suggests that turbid waters are brownish, irrespective of the nature of the matter predominantly defining the water turbidity: it could equally be organic or inorganic in nature. At the same time, waters perceived by the human eye as bluish are only slightly turbid. However, if rather transparent waters contain *doc* they will display, depending on the actual value of C_{doc}, greenish to even brownish colour.

Regarding the behaviour of the other radiometric parameter, colour purity *p*, our numerical simulations indicate that it attains 82% if $C_{chl} = C_{sm} = C_{doc} = 0$, which is in agreement with observed values of *p* for pure oceanic waters (Jerlov, 1976).

We now consider three groups of CPA combinations as displayed in Table 3.5, that largely encompass the multitude of concentrations occurring in natural inland and coastal waters, for a discussion of *p*.

The modelling results indicate that in a two-component aquatic optical system (i.e. either water + *sm* or water + *chl* or water + *doc*) the increase in any admixture or basic component firstly reduces *p* rapidly. This drop is especially drastic for the option water + *sm* (Fig. 3.10), whereas for chlorophyll and particularly dissolved

Table 3.4. Combinations of $C_{chl}/C_{sm}/C_{doc}$ resulting in λ_{dom} restricted to the spectral interval 470–500 nm. Also colour purity p is given.

C_{chl} (μg/l)	C_{sm} (mg/l)	C_{doc} (mg C/l)	λ_{dom} (nm)	p (%)
0	0	0	472	78
0	0	0.5	478	51
0	0	1.0	479	43
0	0	1.5	478	40
0	0	2.0	480	36
0	0	2.5	480	34
0	0	3.0	480	32
0	0	3.5	480	31
0	0	4.0	480	30
0	0	4.5	480	29
0	0	5.0	480	28
0	0	5.5	480	27
0	0	6.0	480	27
0	0	6.5	480	26
0	0	7.0	480	26
0	0	7.5	480	25
0	0	8.0	480	25
0	0	8.5	480	25
0	0	9.0	480	24
0	0	9.5	480	24
0	0	10.0	480	24
0	0.5	0	493	21
1	0	0	480	47
1	0	0.5	485	28
1	0	1	488	20
1	0	1.5	490	15
1	0	2.0	492	12
1	0	2.5	494	10
1	0	3.0	496	8
1	0	3.5	499	7
1	0.5	0	498	16
2	0	0	485	30
2	0	0.5	490	18
2	0	1	495	12
3	0	0	489	21
3	0	0.5	496	13
4	0	0	494	15
5	0	0	500	11

organics it proves to be less steep and less pronounced. However, a further increase in, respectively, either *sm* or *chl* results in a slow recovery of *p*. For *doc* it is different: its enhancement leads to a steady but slow reduction of *p*, and at abnormally high C_{doc} (1000 mg C/l), *p* becomes as low as 0.18 and even 0.10 if CPA cross-sections for

Table 3.5. Variations in C_{chl} (μg/l), C_{sm} (mg/l) and C_{doc} (mg C/l) used for the simulation of p dependence on CPA concentrations.

Case number	Discrete sequence of values of			Quasi-continuous sequence of values of		
	C_{chl}	C_{sm}	C_{doc}	$\Delta C_{chl} = 0.1$	$\Delta C_{sm} = 0.1$	$\Delta C_{doc} = 0.5$
1	0; 1.0; 2.0; 5.0; 10.0; 20.0		0; 0.1; 0.5; 1.0; 5.0; 10.0		0–20.0	
2		0.1; 0.5; 1.0; 5.0; 10.0	0; 0.1; 0.5; 1.0; 5.0; 10.0	0–20.0		
3	0; 0.1; 0.5; 1.0; 5.0; 10.0; 20.0	0; 0.1 0.5; 1.0; 5.0; 5.0; 10.0				0–10.0

Lake Ontario and Lake Ladoga are applied respectively (see Table 3.4 for examples). However, as soon as a third CPA is introduced, the aforementioned sharp decrease in p at low concentrations becomes even steeper with a minimum in p occurring at lower concentrations of the basic component. The absolute minimum in $p(p_{min} \cong 5\%)$ is reached for chlorophyll at $C_{doc} \geq 5$ mg C/l.

When the concentration of the third component exceeds a threshold of five concentration units, the p-minimum is followed by a slow rise in p depending on the basic component concentration (Fig. 3.10). These variations in p become more accentuated when a fourth component (i.e., respectively, either doc or sm or chl) is gradually mixed into the water body. Generally, the transition from the drop to the increase in p occurs within such a narrow range of the basic component concentrations (and these are so small in absolute values) that in reality p starts

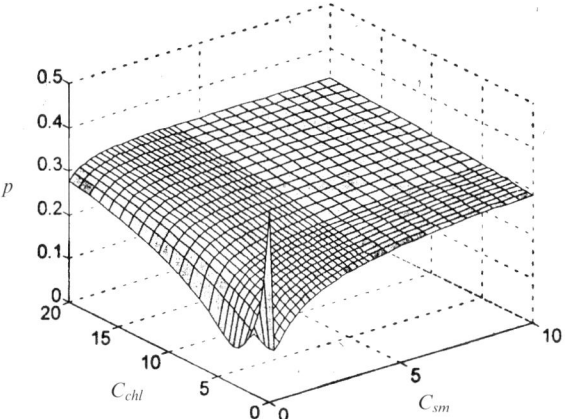

Figure 3.10. Dependence of colour purity p on variations in C_{chl} (μg/l) and C_{sm} (mg/l) at $C_{doc} = 1$ mg C/l.

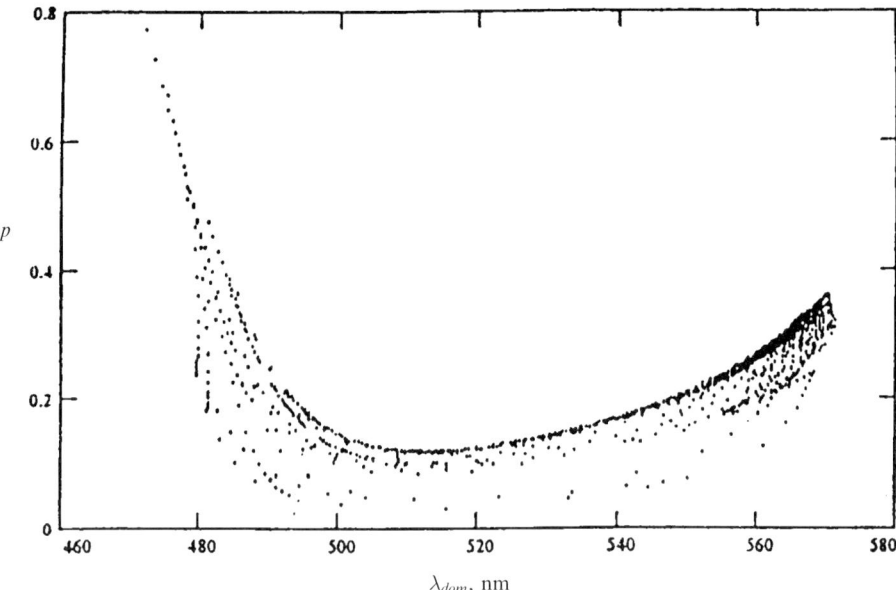

Figure 3.11. Interdependence of simulated p and λ_{dom} values for all CPA combinations specified in Tables 3.1 and 3.5.

increasing immediately when the basic component exceeds several tenths of the respective concentration unit. It could, nevertheless, be pointed out that p_{min} slowly increases with the increasing concentrations of the third and fourth components.

The above p variations resemble the asymptotic pattern of λ_{dom}. But unlike λ_{dom}, the limit value of $p(p_{asymp})$ does not remain the same for different CPA combinations. For those specified in Table 3.5, p_{asymp} lies between 0.35 and 0.37. But, at abnormally high concentrations of all CPAs, p_{asymp} varies from 0.18 to 0.42 and 0.10 to 0.48 for the optical models of Lakes Ladoga and Ontario, respectively (see Table 3.2).

A joint analysis of λ_{dom} and p (Fig. 3.11) reveals that for the spectral region $\leq 480\,$nm high values of p are always associated with low λ_{dom} and, hence, with simultaneously low CPA concentrations. The scatter is due to different CPA combinations, the highest values of p appearing when water is only slightly loaded with either *chl* or *doc* and when C_{sm} is nearly zero. The lowest p values in this spectral zone exist for simultaneously moderate amounts of *chl* and *dom*. The minimum variation in p versus λ_{dom} for all possible combinations of CPAs occurs in the 500–540 nm interval. The lowest values of p occurring at each λ_{dom} belong to C_{sm} far below 1 mg/l, $C_{chl} \approx 1\,\mu$g/l and $C_{doc} \geq 5\,$mg C/l.

Finally, for $\lambda_{dom} \geq 540\,$nm p increases with λ_{dom} up to 0.38 at 575 nm. The lowest p values belong to waters with high *doc* levels and only marginal *chl* and *sm* concentrations. This brings us to a rather interesting conclusion, implicitly already mentioned above, that low p values are not necessarily due to turbid waters with high concentrations of suspended matter but rather are a result of high *doc* levels at fairly low *chl* and *sm* concentrations.

In addition, our findings imply that a green water colour is not restricted to waters heavily loaded with *chl*, but is also found when they are almost totally devoid of *chl* but contain moderate amounts of *doc* and small quantities of *sm*. This phenomenon can be encountered, for instance, in strictly oligotrophic alpine lakes prior to the onset of the phytoplankton vernal development: when viewed from above, these lakes display a bright turquoise colour owing to fine terrigenous/ inorganic matter brought with melting ice/snow and runoff. Recently, Claustre *et al.* (2002) also reported that desert dust makes the oligotrophic waters of the Mediterranean Sea greener than would result from their phytoplankton content alone. Interestingly, Zhang *et al.* (1998) showed by simulations that air bubbles as additional light scatterers can also bring a greenish hue to the water. On the other hand, waters with brownish colour should not necessarily be expected to have high levels of *doc*. They can just be very turbid owing to high concentrations of *chl* and *sm*. This potential ambiguity in attributing some specific CPA composition to an actually observed colour persists almost throughout the whole visible region of the radiation impinging upon the water surface. Without the knowledge of the full spectrum, a separation of CPAs by remote sensing is not unique/unambiguous.

3.1.2 Influence due to inelastic (Raman) scattering

Until now all simulations neglected trans-spectral processes. As the simple simulations were close to observations in many cases, we can expect a major contribution by these trans-spectral processes only for certain CPA concentrations. Firstly, we look at Raman scattering.

Our simulations of the impact of inelastic scattering on radiometric water colour are taken from Pozdnyakov *et al.* (2002a). Following the methodological approach given in Sections 1.1 and 3.1.1, our calculations of the volume reflectance component due to Raman scattering, R_r, are based on eq. (1.36). The input parameters required for the calculation of spectra of R_r include b_r, B_r, μ_d, μ_u, and E_d. $b_r(\lambda)$ has been determined by many workers (for references see Barlett *et al.*, 1998). It lies in the range $(2.7 \pm 0.2) \times 10^{-4}\,\mathrm{m}^{-1}$ at $\lambda = 488\,\mathrm{nm}$. $b_r(\lambda)$ is proportional to λ^{-5} (although it is slightly different for the incident and the Raman scattering wavelengths). The backscattering probability for elastic and inelastic scattering is generally assumed (Marshall and Smith, 1990) to be 0.5 owing to the known shape of the Rayleigh and the isotropic phase functions. $\mu_d(-0, \theta'_0, \lambda)$ was calculated following Berwald *et al.* (1998), and μ_u was assumed to be 0.5 in view of the shape of the Raman scattering phase function.

The incident irradiance was simulated in the spectral range 310–690 nm for cloudless sky conditions using the model suggested by Gregg and Carder (1990). The extension into the ultraviolet is needed due to the Raman frequency shift of

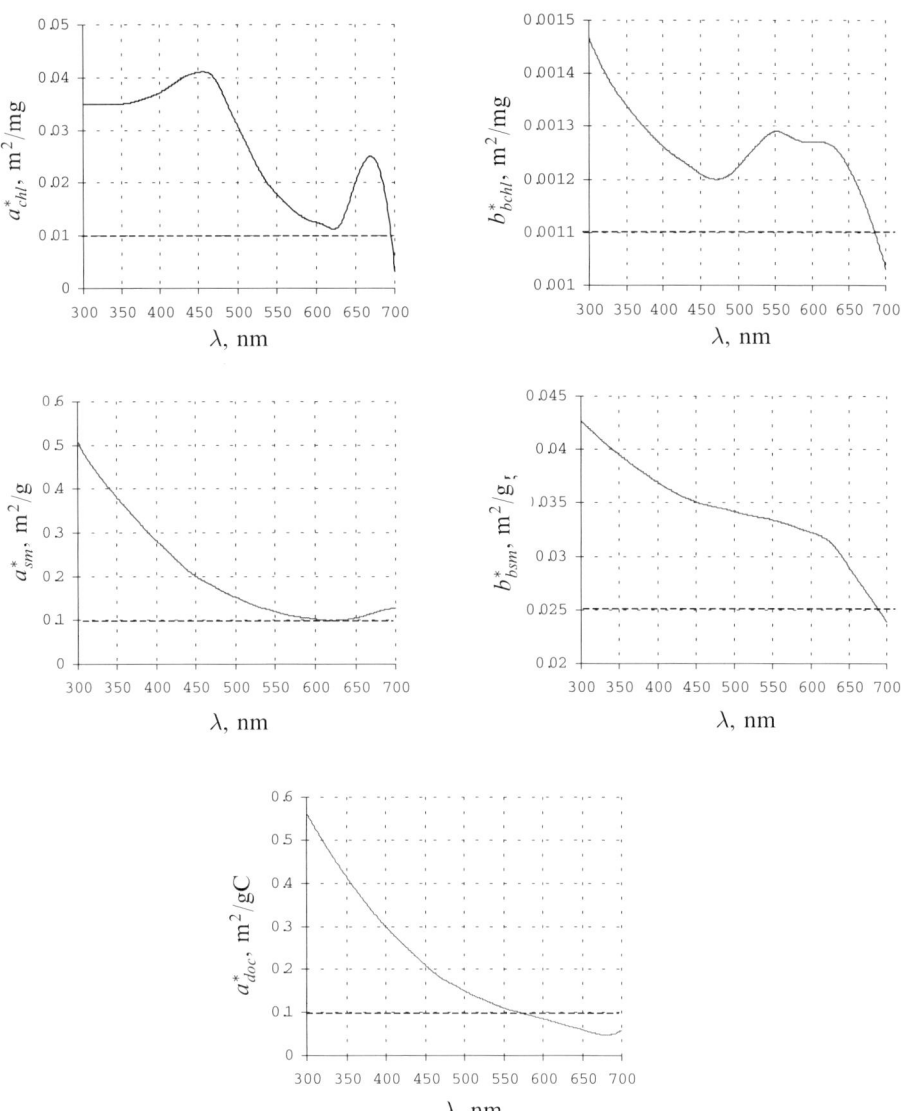

Figure 3.12. Specific cross-section spectra for the hydro-optical model of Lake Ladoga waters with an extension to 310 nm.

about 75 nm leading to emissions at $\lambda \geq 400$ nm stimulated by photons with $\lambda \geq 325$ nm (Sugihara *et al.*, 1984).

To calculate K_d (see eq. (1.29)), F (eventually F_w) is needed as a function of θ_0 and λ; here it was taken from Baker and Smith (1997) for cloudless conditions.

In addition, the backscattering probability for phytoplankton (B_{chl}), suspended minerals (B_{sm}), and water molecules (B_w) needs to be specified to obtain the direct

Table 3.6. Spectral values of the CPA cross-sections in the range 310–410 nm.

λ, nm	a_{chl}^*, m^2 mg^{-1}	a_{sm}^*, m^2 g^{-1}	a_{doc}^*, m^2(g C)$^{-1}$	$b_{b_{chl}}^*$, m^2 mg^{-1}	$b_{b_{sm}}^*$, m^2 g^{-1}
310	0.0350	0.4750	0.5300	0.001 435	0.0420
330	0.0350	0.4250	0.4700	0.001 380	0.0407
350	0.0350	0.3800	0.4150	0.001 340	0.0395
370	0.0355	0.3400	0.3650	0.001 305	0.0384
390	0.0365	0.3000	0.3200	0.001 275	0.0374
410	0.0380	0.2650	0.2800	0.001 250	0.0365

and diffuse light attenuation coefficients (eqs (1.26) and (1.27)). For water, B_w is known to be 0.5 (Morel, 1980). For phytoplankton, B_{chl} is reported to be in the range 0.008–0.018 (Siegel *et al.*, 1997). We used 0.011 as a value recommended by Bukata *et al.*, (1995). Data on B_{sm} are scarce. In our simulations B_{sm} was taken to be equal to 0.08 as it was determined for Lake Ontario waters (Bukata *et al.*, 1995), which is consistent with the relevant laboratory estimations by Vohen (1997). Although it is a rough approximation in the case of phytoplankton, and, perhaps, also suspended minerals, B values were assumed to be wavelength independent.

Finally, for matching the entire spectral range of the incident radiation (i.e. 310–690 nm), the hydro-optical model employed (basically established for 410 nm $\leq \lambda \leq$ 690 nm) was extended to 310 nm (Fig. 3.12, Table 3.6). In view of the capacity of phytoplankton to protect themselves against UV radiation, and accounting for the arguments given by Kondratyev and Pozdnyakov (1996), the absorption cross-sections in the UV were prescribed as follows: $a_{chl}^*(\lambda)$ was assumed to decrease symmetrically (with respect to the band maximum at $\sim$450 nm) until $\lambda = 360$ nm; in the spectral region 360–310 nm, it was held constant at 0.035 m^2/mg. $a_{doc}(\lambda)$ was extrapolated into the short wavelength region following the exponential law inherent in the visible part of the spectrum (Kirk, 1976). Extrapolations of $a_{sm}(\lambda)$, $b_{b_{chl}}$, and $b_{b_{sm}}$ were conducted according to the power law suggested in Bukata *et al.* (1995).

Values of spectral absorption and scattering coefficients, a_w and b_w, for pure water, were adopted from the combined optical model of pure water described in Section 2.2.1.

Only the nadir (i.e. strictly vertical) view was considered in these simulations. Calculations were performed for a calm (flat) water surface and the following depth-independent concentrations of CPAs: ($C_{chl}(\mu g/l) = 0.0, 0.5, 1.0, 2.0, 3.0, 4.0, 5.0, 15.0$; $C_{sm}(mg/l) = 0.01, 0.5, 1.0, 2.0, 3.0, 4.0, 5.0$; $C_{doc}(mg\,C/l) = 0.0, 0.5, 1.0, 2.0, 5.0, 10.0$. These sets of the CPA concentrations embrace most conditions in natural waters from oligotrophic to meso-eutrophic (Petrova, 1990). All calculations were conducted for solar zenith angles $\theta_0 = 0°, 5°, 10°, 30°, 40°$, and $50°$.

Our computational results on spectral variations of the upwelling irradiance, $E_u^r(-0, \lambda)$, arising from pure water Raman scattering ($C_{chl} = C_{sm} = C_{doc} = 0$), indicate that $E_u^r(-0, \lambda)$ increases with λ reaching a maximum at $\sim$430–470 nm which is then followed by a steep fall-off at longer wavelengths. Haltrin and Kattawar (1993) reported a very similar result. This pattern of spectral variations of $E_u^r(-0, \lambda)$ is mostly due to a concerted impact of three factors: (a) the incident radiation, strongly reduced in the short wavelength region, shows a rather broad maximum at about 480–490 nm, and a fall-off at $\lambda > 490$ nm, (b) b_{br} is proportional to $\sim \lambda^{-5}$, and (c) attenuation coefficients K_u, K_d increase for $\lambda > 490$ nm.

However, the relative contribution of Raman scattering to the water-leaving radiance exhibits a different behaviour: it progressively increases with λ, and reaches about 20% at $\lambda \geq 650$ nm. This value is very close to the 19–21% reported in the literature (Waters, 1995). According to Gordon (1999), Raman scattering in pure water contributes about 25% to $L_u(\lambda, 0)$ and $E_u(\lambda, -0)$ for $\lambda \geq 480$ nm.

Calculations for varying solar zenith angle θ_0 give a weak growth of the Raman scattering contribution to $E_u(-0, \lambda)$ with increasing θ_0 for $\lambda \leq 520$ nm. But, the proportion of $E_u^r(-0, \lambda)$ in the total upwelling radiance decreases with increasing θ_0 in the yellow to red portions of the spectrum. Both results agree qualitatively and also quantitatively well with those reported by Waters (1995) and Gordon (1999).

The dependence of R_r on sun zenith angle θ_0 described above is due to the phase function of bulk water and the increase of the photon path length in the upper layers of the water column, and hence by an increase in the photon absorption probability for $\lambda < 400$ nm and increasing R_r in the range yellow–red.

According to Gordon (1999), the absolute contribution of Raman scattering to $L_u(\lambda, -0)$ and $E_u(\lambda, -0)$ is only weakly dependent on the elastic-scattering properties of the particles. Although, of course, $L_u(\lambda, -0)$ is a strong function of the back-scattering probability of suspended particulate matter. Therefore, the Raman contribution to $L_u(\lambda, -0)$ for $C_{chl} \leq 1$ µg/l is still approximately 8% at the wavelengths of interest in ocean colour remote sensing. Thus it cannot be ignored when solving the inverse problem, i.e., when retrieving *chl* concentrations from upwelling radiances. However, this holds only for clear (Case I) oceanic waters.

Returning to Case II waters, our simulations revealed that the addition of either phytoplankton or suspended minerals or dissolved organics to pure water leads to a substantial decrease in $E_u^r(-0, \lambda)$ at wavelengths > 500 nm. Waters (1995) reports a similar behaviour of $E_u^r(-0, \lambda)$ with increasing phytoplankton concentration in a two-component (pure water plus phytoplankton) system. This effect is due to the ability of the above CPAs to absorb light in the short wavelength region, and it is most accentuated in the case of dissolved organics whose absorptivity increases exponentially with decreasing λ.

Figure 3.13 illustrates the relevant spectral variations of total volume reflectance, R_{tot} (i.e. volume reflectance composed of volume reflectance due to elastic scattering, R_{es} and R_r) at $\theta_0 = 10°$ for three modelling options: pure water plus phytoplankton (Fig. 3.13(a)), pure water plus suspended minerals (Fig. 3.13(b)), and pure water plus dissolved organics (Fig. 3.13(c)). Increasing amounts of each of these constituents of natural water reduce the resultant volume reflectance down to the R_{es} value (Table 3.7), thus indicating that even at high solar zenith angles the contribution of Raman scattering to the volume reflectance in moderately to

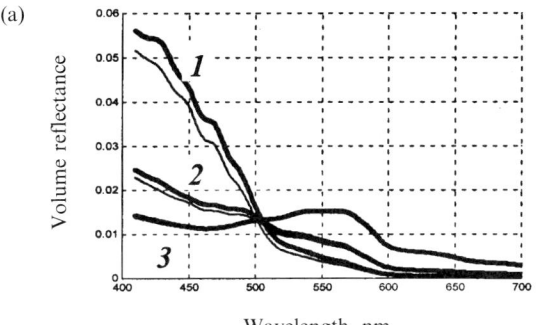

Figure 3.13. Impact of Raman scattering by water molecules on the total volume reflectance for a number of hydro-optical options: (a) $C_{sm} = C_{doc} = 0$, $\mathbf{1} - C_{chl} = 0$, $\mathbf{2} - C_{chl} = 1.0\,\mu g/l$, $\mathbf{3} - C_{chl} = 5.0\,\mu g/l$; (b) $C_{chl} = C_{doc} = 0$, $\mathbf{1} - C_{sm} = 0$, $\mathbf{2} - C_{sm} = 0.1\,mg/l$, $\mathbf{3} - C_{sm} = 1.0\,mg/l$; (c) $C_{sm} = C_{chl} = 0$, $\mathbf{1} - C_{doc} = 0$, $\mathbf{2} - C_{doc} = 0.1\,mg\,C/l$, $\mathbf{3} - C_{doc} = 1.0\,mg\,C/l$. Solar zenith angle $\theta_0 = 10°$. Thin lines refer to R_{es} only.

Table 3.7. Relative contribution to total spectral volume reflectance $R_{tot}(\lambda, -0)$ by water Raman scattering for a variety of hydro-optical options at $\theta_0 = 10°$. Concentrations of chlorophyll, suspended minerals, and dissolved organics are given in µg/l, mg/l, and mg C/l, respectively. Fluorescence effects due to chlorophyll and dissolved organics are neglected.

λ, nm	Options					
	water + chl		water + sm		water + doc	
	$C_{sm} = C_{doc} = 0$ $C_{chl} = 1$	$C_{chl} = 5$	$C_{chl} = C_{doc} = 0$ $C_{sm} = 0.1$	$C_{sm} = 1$	$C_{chl} = C_{sm} = 0$ $C_{doc} = 0.1$	$C_{doc} = 1$
400	0.08	0.04	0.05	0.01	0.01	0.01
500	0.07	0.02	0.04	0.00	0.11	0.04
650	0.07	0.02	0.04	0.00	0.26	0.22

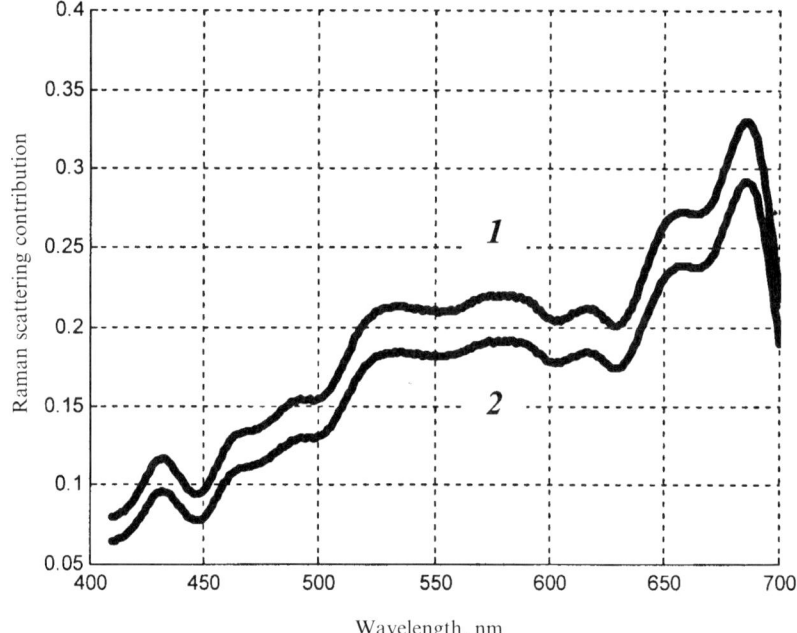

Figure 3.14. Raman scattering contribution in pure water to the total volume reflectance $R(\lambda, -0)$ for two zenith angles: *1* – $\theta_0 = 10°$; *2* – $\theta_0 = 45°$.

strongly turbid/absorbing waters is small at all wavelengths. At the same time, in accordance with the above, this mechanism of sunlight interaction with the water molecules can be consequential for the formation of colour of Case I waters.

The relative contribution of R_r to the total volume reflectance, R_{tot}, slightly decreases with increasing θ_0, mostly as a result of two causes. Firstly, it is known (Gordon, 1989) that $R_{es}(-0, \lambda)$ increases with increasing θ_0. Secondly, as we saw, the

Table 3.8. Relative contribution to total spectral volume reflectance R_{tot} by water Raman scattering for two sun zenith angles θ_0 in the case of pure water ($C_{chl} = C_{sm} = C_{doc} = 0$).

λ, nm	θ_0	
	10°	45°
400	0.09	0.07
500	0.15	0.13
650	0.26	0.23

contribution of $E_u^r(-0, \lambda)$ to $E_u(-0, \lambda)$ decreases with increasing θ_0. The spectral variation of this dependence is illustrated both in Fig. 3.14 and Table 3.8.

Overall, our calculations indicate that Raman scattering should only have a small impact on the radiometric colour characteristics of non-Case I waters.

3.1.3   Fluorescence by chlorophyll and dissolved organics

The fluorescence models (eqs (1.38) and (1.40)) require, in addition to the adopted hydro-optical model, the following input data: downwelling irradiance $E_d(\lambda)$, the fluorescence yield for phytoplankton η_{chl} and dissolved organics η_{doc}, the standard deviation of the fluorescence band ($\sigma_{chl}, \sigma_{doc}$), and the wavelength of maximum fluorescence $\lambda_{0\,em}$ both for chlorophyll and dissolved organics. In our simulations (Pozdnyakov *et al.*, 2002a), $E_d(\lambda)$ was calculated as in Section 3.1.2.

The phytoplankton fluorescence yield η_{chl} was assessed by several workers (Gordon, 1979; Kondratyev *et al.*, 1999) and was shown to be strongly dependent on the taxonomic composition of phytoplankton, incident radiation intensity, availability of nutrients, and some other factors. Accordingly, η_{chl} varies in a wide range from about 1% to 12% with the most common numbers from 1% to 3% (under conditions of maximum *photoinhibition*). For instance, Babin *et al.* (1996) investigated the latitudinal and/or seasonal distribution of η_{chl} in ocean waters. They conclude that η_{chl} varies between 2.7% and 4.5% with maximum values in temperate latitudes. However, according to their estimations, the actual η_{chl} might be as low as 1% due to a combined action of photoprotectant and nutrient limitation effects. Alongside these data, there are indications that η_{chl} can be as small as 0.3% (Fischer and Kronfeld, 1990). The available data on η_{chl} point to a wavelength independence of this physical quantity, which is also a function of the algal species. In view of all these uncertainties, we have explored, for comparison reasons, two options with $\eta_{chl} = 0.7\%$ and 3%, neglecting a wavelength dependence over the entire PAR region.

Recently, Maritorena *et al.* (2000) assessed the values of η_{chl} for oceanic phytoplankton in their natural habitats in the central Pacific. At chlorophyll concentrations from 0.04 to 3 μg/l and solar noon, the vertical profiles of η_{chl} were strongly structured, generally varying at depth between 2% and 3% with maximal values up to 5–6% and only $\approx$1% in near-surface waters. This decrease of η_{chl} in the top layer is thought to be a result of the suppression of photosynthetic activity at high levels of radiation. In marine oligotrophic waters, η_{chl} is always low and increases only by a factor of 2 from surface to depth. In eutrophic marine waters, it remains low at the surface, but it increases by a factor of 5 and even 10 at depth. For remote sensing applications, only η_{chl} in the upper layers is relevant, and, judging from these results, will not exceed about 1%, at least in oceanic waters and at high sun elevation.

Although being even rarer than for η_{chl}, the data on the fluorescence yield of dissolved organics, η_{doc}, indicate strong dependence on λ. Based on some river water data, its values vary from about 0.5% to 1.0% at $\lambda = 250$ nm to 0.5% to 1.9% at $\lambda = 470$ nm with a maximum ranging from 0.7% to 2.8% at $\lambda = 390$ nm (Green and Blough, 1994). Here we use the following spectral values of η_{doc}:

λ, nm	310	330	350	370	390	410	430	450	470
η_{doc}, in %	1.25	1.75	2.30	2.45	2.70	2.60	2.30	2.10	0

Chlorophyll-a fluoresces $in\ vivo$ from 660 to 760 nm with the maximum at about 685 nm (Gordon, 1979). Owing to an extremely complex and water body-dependent chemical composition, dissolved organics exhibit less rigid boundaries of their fluorescence emission spectra, ranging from about 400 nm to 620–660 nm and peaking at 490–510 nm (Vodacek et al., 1994). However, Coble and Brophy (1996) found that the dissolved organics fluorescence maximum is located at somewhat shorter wavelengths. In conformity with these data on fluorescence yields and emission spectra we represented the emission band by a Gaussian function centred at 685 nm for chlorophyll and at 490 nm for dissolved organics with the widths at half-maximum of 25 nm and 100 nm, respectively.

The Lake Ladoga model was used in its spectrally extended form as described above. In applying eqs (1.38) and (1.40), it was assumed that $a_f \equiv a_{chl}$ or a_{doc}. If in the case of chlorophyll this assumption is easily justifiable, it appears less certain for dissolved organics. Indeed, there are indications (Kondratyev et al., 1999) that only a certain proportion of dissolved organics fluoresces. However, being unaware of any data on this score, we assumed identity between a_f and a_{doc}. The integration in eqs (1.38) and (1.40) was performed over the entire spectral region from 310 to 690 nm.

The use of both fluorescence models (eqs (1.38) and (1.40)) for the calculation of relative contributions of $R^f(\lambda_{em})$ to the total volume reflectance reproduced earlier results for flourophor-specific fluorescence models. Figures 3.15 and 3.16 illustrate, for a variety of hydro-optical conditions, contributions by fluorescence of chlorophyll and dissolved organics, respectively. As in earlier studies, the spectral region with notable contributions from fluorescence by chlorophyll lies above 660 nm and between 430 and 660 nm for fluorescence by dissolved organics. Culver and Perry (1997) report that solar-induced chlorophyll-a fluorescence accounts for 10% to 40% of the total upwelling spectral irradiance at the surface. Vodacek et al. (1994) report that for dissolved organics the fluorescence contribution to the reflectance in

(a)

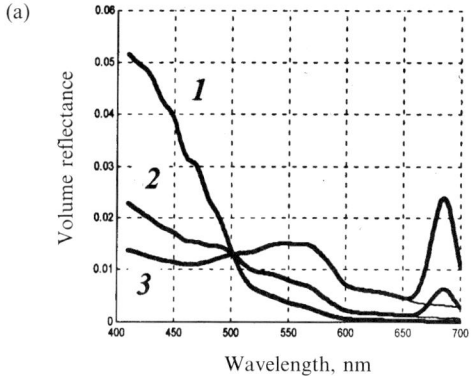

(b)

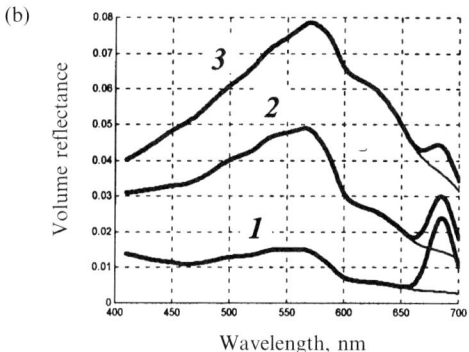

(c)

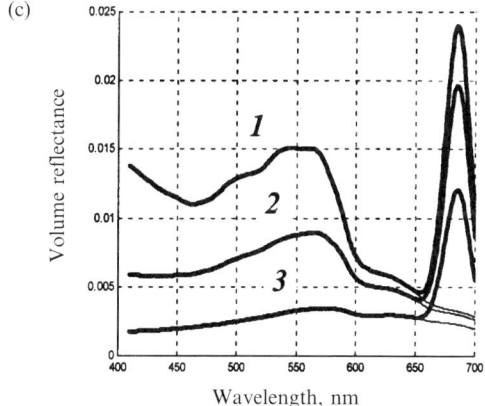

Figure 3.15. Chlorophyll fluorescence $(\eta_{chl} = 3\%)$ impact on spectral volume reflectance $R(\lambda, -0)$ for a number of hydro-optical options: (a) $C_{sm} = C_{doc} = 0$, $\boldsymbol{1} - C_{chl} = 0$, $\boldsymbol{2} - C_{chl} = 1.0\,\mu g/l$, $\boldsymbol{3} - C_{chl} = 5.0\,\mu g/l$; (b) $C_{chl} = 5\,\mu g/l$, $C_{doc} = 0$, $\boldsymbol{1} - C_{sm} = 0$, $\boldsymbol{2} - C_{sm} = 1.0\,mg/l$, $\boldsymbol{3} - C_{sm} = 5.0\,mg/l$; (c) $C_{sm} = 0$, $C_{chl} = 5\,\mu g/l$, $\boldsymbol{1} - C_{doc} = 0$, $\boldsymbol{2} - C_{doc} = 1.0\,mg\,C/l$, $\boldsymbol{3} - C_{doc} = 5.0\,mg\,C/l$. Solar zenith angle $\theta_0 = 10°$. Thin lines refer to R_{es} only.

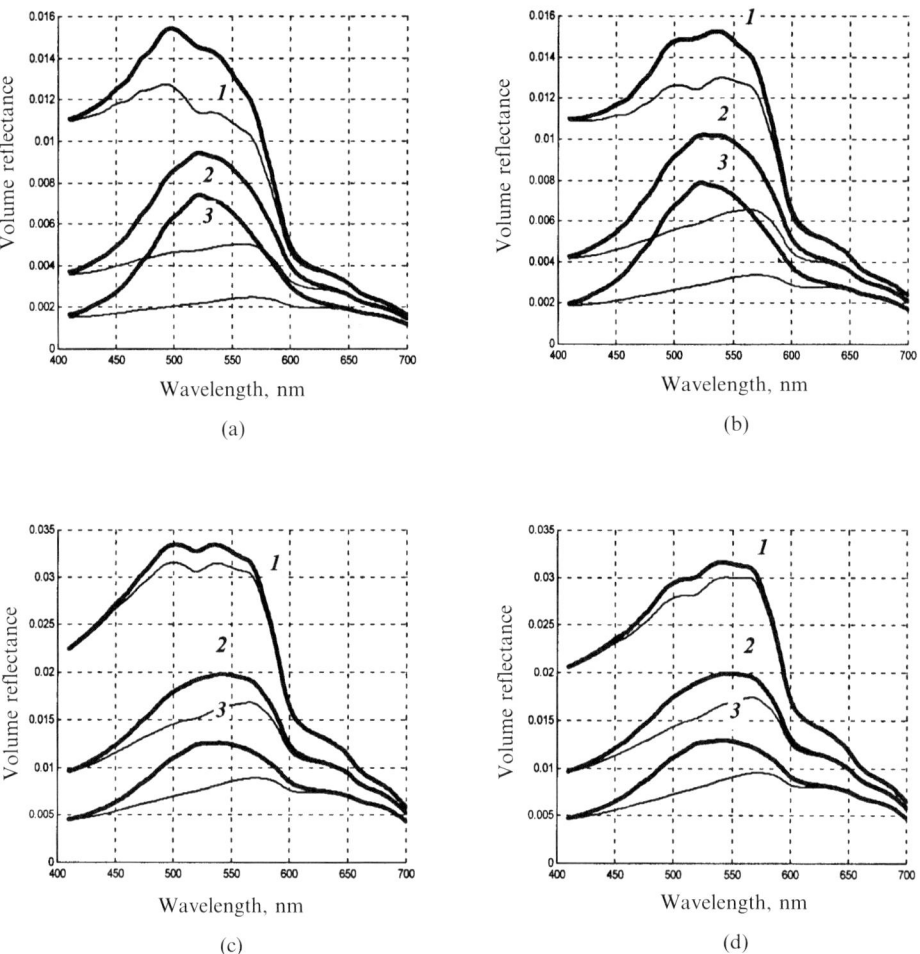

Figure 3.16. Dissolved organics fluorescence impact ($\eta_{doc}(\lambda)$ as tabulated in the text) on the volume reflectance $R(\lambda, -0)$ or a variety of CPA concentration vectors: (a) $C_{chl} = 0.5\,\mu g/l$, $C_{sm} = 0.10\,mg/l$; (b) $C_{chl} = 2.0\,\mu g/l$, $C_{sm} = 0.10\,mg/l$; (c) $C_{chl} = 0.5\,\mu g/l$, $C_{sm} = 0.5\,mg/l$; (d) $C_{chl} = 2.0\,\mu g/l$, $C_{sm} = 0.5\,mg/l$, $C_{doc} = 0.5\,mg\,C/l$ (**1**), $2.0\,mg\,C/l$ (**2**), $5.0\,mg\,C/l$ (**3**). Solar zenith angle $\theta_0 = 10°$. Thin lines refer to R_{es} only.

the blue-green to green spectral region can be as high as 70% for water without suspended sediments and phytoplankton, so called black water.

These two assessments compare well with our results in Fig. 3.17, which also shows the dependence of the fluorescence contribution to total volume reflectance on solar zenith angle. However, the curves in Figs 3.15 and 3.16 clearly indicate a strong sensitivity of R^f (both for chlorophyll and dissolved organics fluorescence) to the presence of absorbing and/or scattering matter: R^f rapidly decreases as CPA concentrations grow. Suspended minerals are especially efficient in damping the

(a)

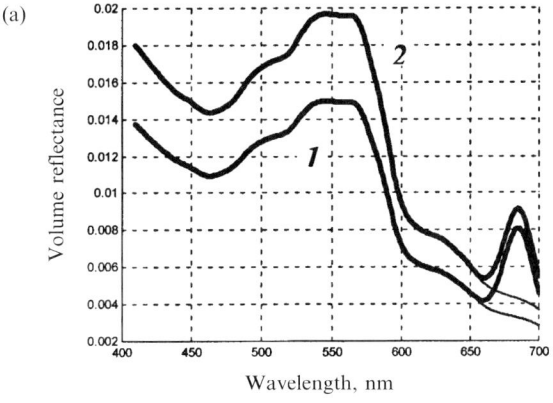

(b)

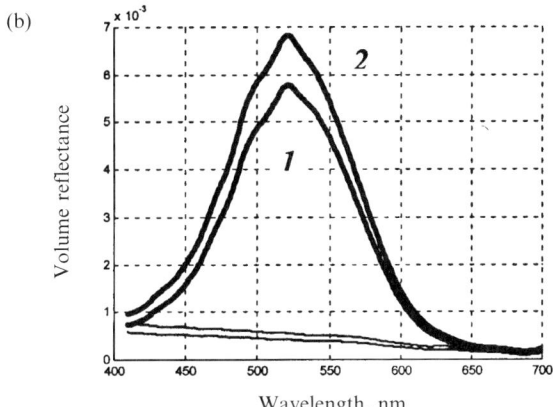

(c)

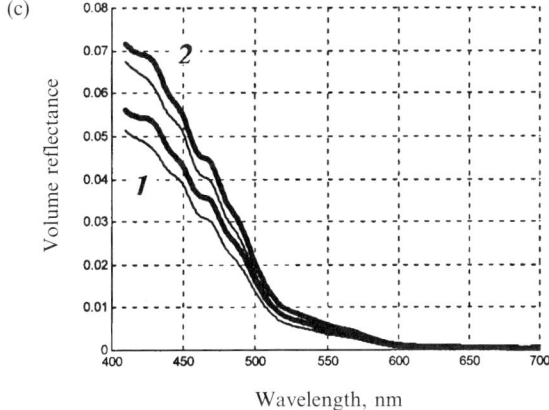

Figure 3.17. The influence of sun zenith angle on the volume reflectance $R(\lambda, -0)$ including fluorescence and water Raman scattering: (a) $C_{chl} = 5\,\mu g/l$, $C_{doc} = C_{sm} = 0$; (b) $C_{sm} = C_{chl} = 0$, $C_{doc} = 5\,mg\,C/l$; (c) $C_{doc} = C_{sm} = C_{chl} = 0$. Thin lines refer to R_{es} only. $\eta_{chl} = 0.7\%$, $\eta_{doc}(\lambda)$ as tabulated in the text. Solar zenith angle $\theta_0 = 5°$ (**1**) and $50°$ (**2**).

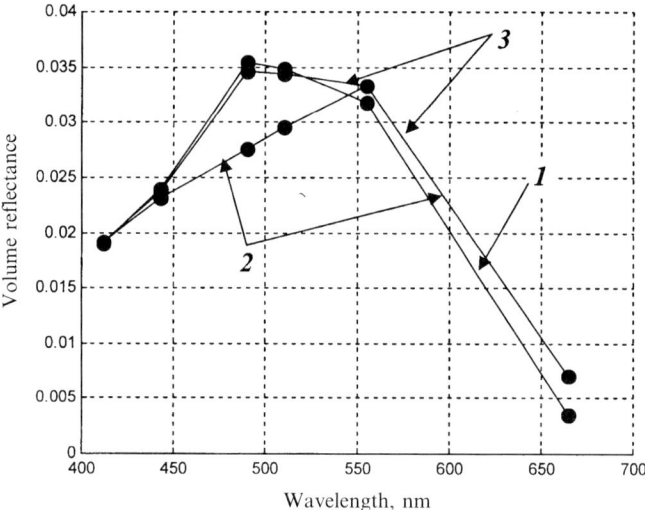

Figure 3.18. Comparison of measured spectral volume reflectance $R(\lambda, -0)$ in the outer part of Saginaw Bay (*1*) with spectra calculated with (*3*) and without (*2*) consideration of fluorescence of dissolved organics.

fluorescence-driven part of volume reflectance. Therefore, the impact of all three trans-spectral processes on upwelling radiance, and thus irradiance will be very low in strongly absorbing/scattering waters. However, it is obviously not the case for *moderately* absorbing/scattering waters.

A fluorescence 'signal' in the spectral distribution of volume reflectance can be detected in waters rich either in phytoplankton or dissolved organics but containing only small amounts of suspended minerals.

We believe that we have an experimental evidence of the impact of fluorescence by dissolved organic matter on $R(\lambda, -0)$. Simultaneous measurements of CPA concentrations and subsurface upwelling and downwelling irradiances at the SeaWiFS wavelengths: 412, 443, 490, 510, 555, 670 nm were conducted in Saginaw Bay of Lake Huron (Pozdnyakov *et al.* 1999). Regarding trophic conditions, Saginaw Bay is generally subdivided into the inner and outer part, the former being eutrophic, the latter oligotrophic. Importantly, being relatively deficient in phytoplankton and suspended minerals, waters in the outer part are rather rich in dissolved organics, and thus constitute favourable conditions for the detection of the fluorescence signal of *doc* in $R(\lambda, -0)$. Indeed, more than 20 determinations of $R(\lambda, -0)$ in the outer part of Saginaw Bay revealed, under varying sun zenith angles ($22° < \theta_0 < 50°$) and wind speed (calm to 5 m/s), a maximum at about 490 nm and an inflection at 555 nm. Figure 3.18 (curve *1*) illustrates such a spectrum for one of the stations. The concentrations of chlorophyll, suspended minerals, and dissolved organics determined *in situ* at this station were 1.13 µg/l, 0.14 mg/l, and 1.14 mg C/l, respectively. The solar zenith angle θ_0 was about 30°.

Several hydro-optical models were tested to simulate the experimental spectrum of $R(\lambda, -0)$. The best fit was attained (Fig. 3.18, curve **2**) with the Ontario Lake model (Bukata *et al.*, 1995). However, even this model failed to reproduce the main maximum at $\sim 490\,\text{nm}$. The inclusion of fluorescence by dissolved organics and chlorophyll, using the above methodology and η_{doc} as in the input data section, as well as $\eta_{chl} = 0.7\%$, resulted in a good match between the measured and simulated $R(\lambda, -0)$ spectra (Pozdnyakov *et al.*, 1999), giving evidence of the impact of fluorescence by *doc* on water leaving radiance in natural waters.

This is further confirmed by data obtained during the same field experiment: the maximum in $R(\lambda, -0)$ at about 490 nm gradually declines and turns into an inflectation for stations nearer to the inner part of the bay where the water is more abundant in chlorophyll and suspended minerals. Within the inner bay, spectra of $R(\lambda, -0)$ do not display any specific feature at 490 nm. Disappearing completely, this feature gives way to very low (1% or less) values of $R(\lambda, -0)$ in the spectral region 400–530 nm with a maximum of $R(\lambda, -0) \cong 3\%$ shifted to 560 nm.

Given the typical concentrations of CPAs observed in the inner part of the bay ($C_{chl} \geq 10\,\mu\text{g/l}$, $C_{sm} = 2\text{–}3\,\text{mg/l}$, $C_{doc} \cong 3\,\text{mg C/l}$), the disappearance of the spectral feature at 490 nm in the simultaneously measured $R(\lambda, -0)$ spectra corroborates our forward modelling results. Indeed, as it was shown above, the fluorescence signal by dissolved organics is most pronounced in the spectrum of $R(\lambda, -0)$ when the water is nearly devoid of chlorophyll and suspended minerals. Enrichment of water with chlorophyll and suspended minerals (at a given C_{doc}) results in a rapid damping of the fluorescence contribution to $R(\lambda, -0)$ due to dissolved organics (cf. Figs 3.16(a) and 3.16(d)). Thus, with chlorophyll and suspended minerals increasing from the outer to the inner part of the bay, the contribution of dissolved organics fluorescence to $R(\lambda, -0)$ first transforms from a maximum to an inflexion and then disappears completely, becoming masked by the predominant impact of absorption by both chlorophyll and suspended minerals (for more details see Pozdnyakov *et al.*, 1999).

Finally, in support of the suggested interpretation of the Saginaw Bay spectral data and our numerical modelling results, it is worth mentioning that analogous spectral distributions of $R(\lambda, -0)$ exhibiting either a maximum or inflexion at $\sim 490\,\text{nm}$ have been recorded in some small (sometimes, nearly pristine) lakes in northern Ontario in the cases when dissolved organics proved to be relatively high, whereas chlorophyll and suspended minerals were present in low amounts (F. J. Tanis, 1999, personal communication).

In support of the above argumentation, we also mention the documented variance [within the range $0.011\text{–}0.022\,\text{nm}^{-1}$] in the exponent s in eq. (2.1), which is thought (Carder *et al.*, 1999) to implicitly indicate the optical impact of fluorescence of dissolved organics in natural waters: an increase in the exponent s is argued to be an indication of low levels of *doc* fluorescence due to shadowing/quenching/biological depletion, etc. of *doc*.

The signal arising from sun-stimulated fluorescence by phytoplankton chlorophyll was detected spectrometrically in natural waters more than 20 years ago (Neville and Gower, 1977). It was further substantiated by airborne and shipborne observations (Doerffer, 1981; Lin *et al.*, 1984; Gower and Borstad, 1990).

Gower *et al.* (1999) reexamined the signature of the chlorophyll fluorescence peak, and have shown by high-resolution spectrometric measurements that the spectral features in the red part of the spectrum associated with the fluorescence of chlorophyll are affected by absorption of pure water, scattering by suspended particles, absorption by chlorophyll per se, and when applicable, by submerged macrophyta. Also, these findings conform well with our numerical results.

Based on our numerical analyses of the impacts of fluorescence by chlorophyll and dissolved organics on $R(\lambda, -0)$, we consider, finally, the implications of these impacts on the radiometric characteristics of water colour in comparison with the results we obtained, assuming that sun–photon interactions in aquatic media are solely restricted to absorption and backscattering. Remember that these latter results reveal the following major features (Bukata *et al.*, 1995):

(1) Natural waters that contain simultaneously low concentrations of chlorophyll, suspended minerals, and dissolved organics appear blue to turquoise in colour (dominant spectral wavelength λ_{dom} in the range 472–500 nm).
(2) Highly turbid waters (i.e. waters containing high concentrations of chlorophyll and/or suspended minerals) display colours ranging from green to brown (i.e. $\lambda_{dom} > 500$ nm).
(3) Waters with high concentrations of dissolved organics, irrespective of turbidity, are invariably perceived as brownish (λ_{dom} in the range 560–570 nm), unless the concentrations of chlorophyll and suspended minerals are infinitesimally low (in the latter case λ_{dom} remains nearly invariant and equal to ~480 nm provided $C_{doc} \geq 0.5$ mg C/l).
(4) Increasing the content of *all* CPAs in natural waters, either individually or collectively, results in an asymptotic approach to an 'end-point' dominant wavelength at about 572 nm, an 'end-point' colour physically located in the yellow-green region of the visible spectrum.
(5) When one or more CPAs exceed a critical concentration, the spectral purity, p, of the 'end-point' colour asymptotically approaches values in the range 0.35 to 0.45.
(6) The yellow-green 'end-point' colour of all natural waters with increasing CPAs, coupled with a low spectral distinctiveness (low 'end-point' purity, p) results in the perceived brownish 'end-point' colour of all very turbid/absorbing natural water bodies.

The consideration of fluorescence by dissolved organics and phytoplankton chlorophyll leads to substantial departures from the above generalities. In the first place, the aforementioned irresponsiveness of λ_{dom} to increasing amounts of dissolved organics (beyond a certain threshold of C_{doc}) at very low concentrations of chlorophyll (<0.2 µg/l) and suspended minerals (<0.1 mg/l) is lost, at least for $C_{doc} \leq 10$ mg C/l (Figs 3.19 and 3.20, computations for $\theta_0 = 30°$, $\eta_{chl} = 0.7\%$).

In addition, the common 'end-point' colour at high concentrations of CPAs no longer exists in the range $0 < C_{doc} \leq 10$ mg C/l, when the fluorescence impacts are included. Rather, this common 'end-point' colour becomes concentration vector-specific. Importantly, the dominant wavelengths corresponding to these specific

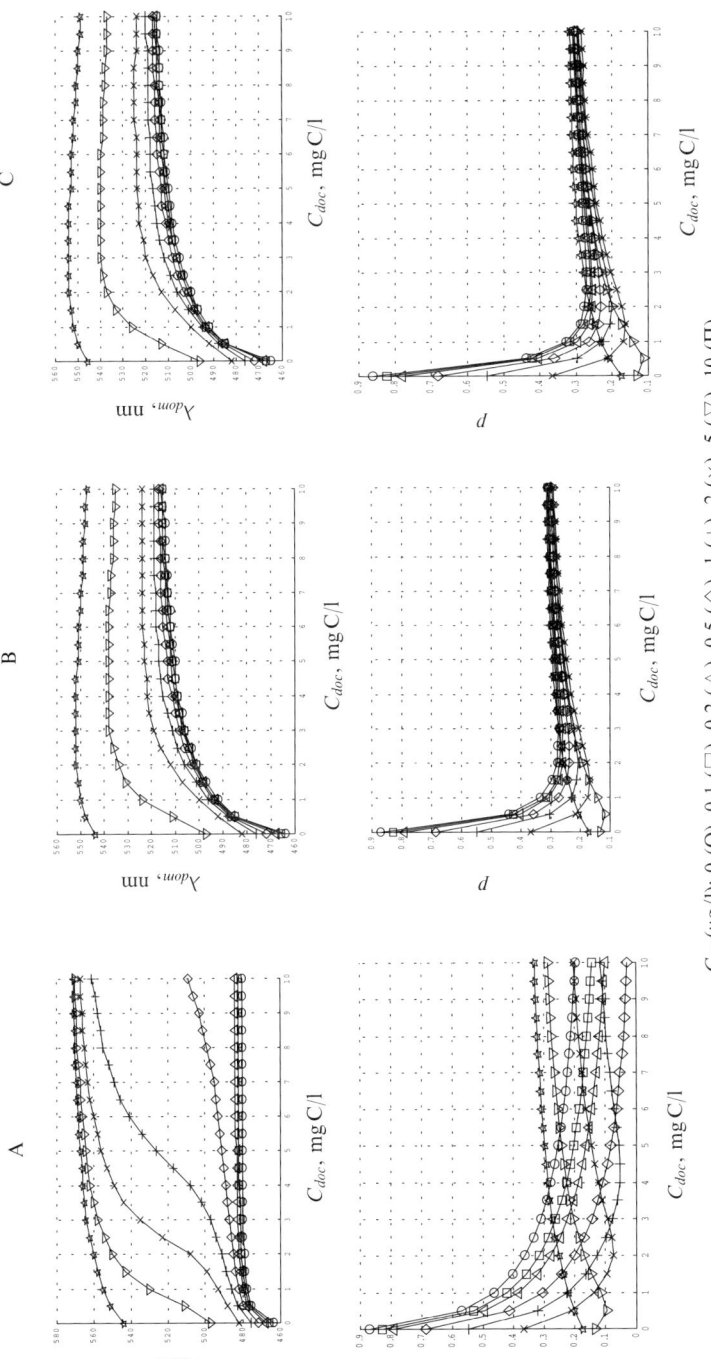

Figure 3.19. Variations of the dominant wavelength λ_{dom} and colour purity p in water devoid of suspended minerals when the impacts due to the fluorescence of dissolved organic matter and chlorophyll are neglected (A) and taken into consideration (B, C). B, only the fluorescence by dissolved organic matter; C, the fluorescence by both dissolved organic matter and chlorophyll. $\eta_{chl} = 0.7\%$, $\eta_{doc}(\lambda)$ as tabulated in the text.

$C_{chl}(\mu g/l)$: 0 (O), 0.1 (□), 0.2 (△), 0.5 (◇), 1 (+), 2 (×), 5 (▽), 10 (Π)

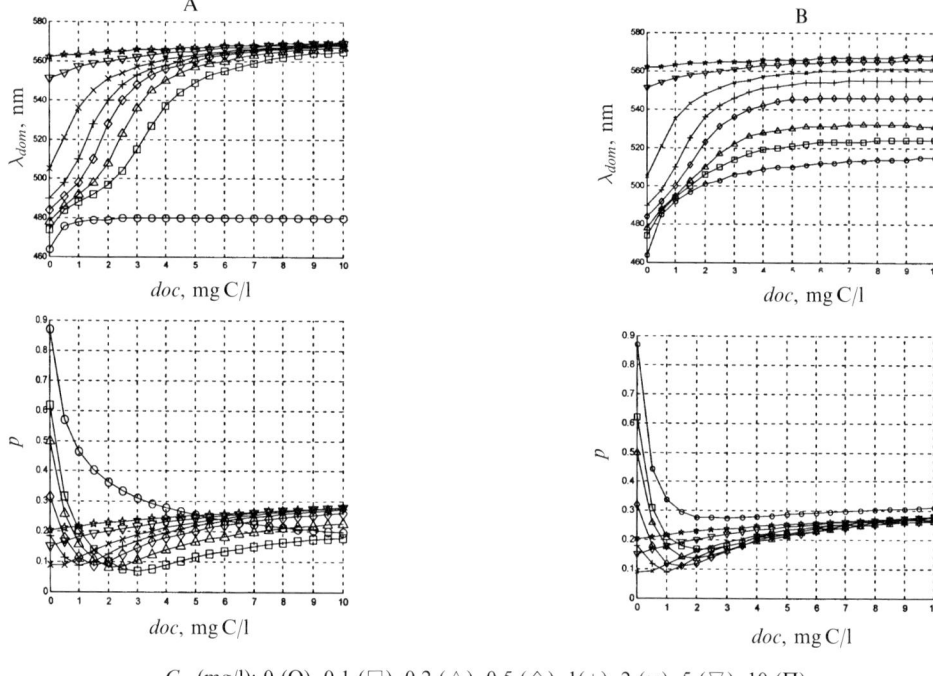

C_{sm}(mg/l): 0 (O), 0.1 ($\square$), 0.2 ($\triangle$), 0.5 ($\Diamond$), 1(+), 2 ($\times$), 5 ($\triangledown$), 10 (Π)

Figure 3.20. Variations of the dominant wavelength λ_{dom} and colour purity p in water devoid of chlorophyll when the fluorescence by dissolved organic matter is neglected (A) or taken into consideration (B). $\eta_{chl} = 0.7\%$, $\eta_{doc}(\lambda)$ as tabulated in the text.

'end-point' colours are generally substantially shorter (unless $C_{sm} \geq 5$ mg/l) than the 'end-point' λ_{dom} in the 'absence' of fluorescence (i.e. 560–570 nm). This holds especially when the water turbidity is controlled exclusively by phytoplankton (i.e. $C_{sm} = 0$), e.g. λ_{dom} of the 'end-point' colour at $C_{chl} = 10$ μg/l is <550 nm as compared to ~570 nm when the fluorescence impact is neglected (Figs. 3.19 and 3.20).

With increasing C_{chl}, the chlorophyll fluorescence causes a more rapid approach to the 'end-point', although at smaller λ_{dom}. Importantly, this shift of the 'end-point' to shorter wavelengths is nearly independent of η_{chl}.

As can be seen from Figs 3.19 and 3.20, in contrast to the behaviour of λ_{dom}, the asymptotic approach of colour purity, p, to values in the range 0.35 to 0.45 (when one or more CPAs is (are) in excess of a critical concentration) becomes more rapid with fluorescence.

In summary, simulations using widely accepted relations between CPA concentrations on the one hand and optical properties and fluorescence on the other hand show that fluorescence mechanisms influence the colour of natural water bodies, especially when the concentration of one of the fluorophors is relatively high whereas the other CPAs show low concentrations. This conclusion is also supported by the aforementioned data from Saginaw Bay. For instance, at a station with $C_{chl} = 1.13\,\mu g/l$, $C_{sm} = 0.14\,mg/l$, $C_{doc} = 1.14\,mg\,C/l$ in the outer part of Saginaw Bay λ_{dom} was equal to 501 nm, but becomes 525 nm after subtraction of the dissolved organics fluorescence signal. As could be anticipated from the above discussion, the colour purity, p, undergoes only minor changes from 0.09 to 0.08 for this concrete case study.

As λ_{dom} varies more strongly at high concentrations of phytoplankton and dissolved organics if fluorescence is taken into account, this will also offer additional information if remote sensing algorithms take fluorescence into account.

3.2   OPTICALLY SHALLOW WATERS: IMPACT OF BOTTOM ALBEDO

It was underscored in Chapter 1, Section 1.5, that in optically thin/shallow waters a certain amount of photons propagating downwards from the air–water interface can be reflected at the bottom instead of ultimately being absorbed by the aquatic medium as generally occurs in optically semi-infinite waters. Therefore, in the case of optically shallow waters, a fraction of the photons reflected at the bottom can reach the water–air interface, so that the resulting subsurface upwelling irradiance $E_{u,tot}(\lambda, -0)$ becomes a sum of two components: $E_{u,wc}(\lambda, -0)$ and $E_{u,bot}(\lambda, -0)$, the upwelling irradiances due to photons which have not interacted or have interacted with the bottom, respectively (see eq. (1.41)).

When discussing the expression for $E_{u,tot}(\lambda, -0)$ (eq. (1.45) in Section 1.5), it was pointed out that for the downward and upward propagating fluxes through the diffuse attenuation coefficient radiation $K_d(\lambda, \theta_0)$, with θ_0 being the sun zenith angle, and $k(\lambda)$ are not equal, and do not remain invariant with depth as a result of changes in the angular distribution of photon fluxes in both directions (Kirk, 1999).

Even under the assumption of isotropic optical properties of the water column, the diffuse attenuation coefficients determining the irradiances, $E_{u,wc}(\lambda, -0)$ and $E_{u,bot}(\lambda, -0)$ are not identical and should be distinguished ($k_1(\lambda)$ and $k_2(\lambda)$). Indeed, since the mean cosine μ_d for the downward propagating sunlight is larger than the mean cosine μ_u for the upward propagating sunlight, it is obvious that $k_1(\lambda)$ should be larger than $K_d(\lambda, \theta_0)$. This excess of $k_1(\lambda)$ over $K_d(\lambda, \theta_0)$ should be especially significant in the imaginary case of 'black' skies, when the atmosphere is absent and the entire incident radiation is composed of direct sun rays. The excess can become small at low sun zenith angles, θ_0, and close to the air–water interface. According to Kirk (1984, 1999), this excess can reach a factor of 2.5. However, as the direct sunlight penetrates deeper into the water column, μ_d shows a smooth decrease as a result of scattering dominating over absorption. At the same time, $k_1(\lambda)$ does not change significantly with depth, so that at lower levels the difference between $k_1(\lambda)$ and $K_d(\lambda)$ shrinks.

However, an entirely different situations exists for the diffuse attenuation coefficient $k_2(\lambda)$ related to the bottom reflected upwelling sunlight. Assuming that the bottom is a Lambertian reflector with an albedo A, the mean cosine for the reflected photons can be set to $\mu_{u,bot} \cong 0.5$, and the reflected flux of photons travelling to the water–air interface should be subject to weaker attenuation as compared to the upwelling radiance caused by backscattering of the water column, i.e. $k_1(\lambda) > k_2(\lambda)$. Consequently, $k_2(\lambda)$ should conform with the following inequality: $k_2(\lambda) < 2K_d(\lambda)$. The problem, however, resides in the principal difficulty in separate measurements of the diffuse attenuation coefficients $k_1(\lambda)$ and $k_2(\lambda)$, whereas the coefficient of diffuse attenuation of downwelling irradiance, $K_d(\lambda)$ is routinely and sufficiently accurately measured, provided the sensor is stably oriented in the horizontal plane.

Resorting to the obviously rough simplification of an identity between $K_d(\lambda)$ and k_1, k_2 for all photon fluxes under consideration, one gets expression (1.46), relating the total volume reflection just beneath the water–air interface to its components and major variables:

$$R_{tot}(\lambda, -0) = R_{wc}(\lambda, -0) + (A(\lambda) - R_{wc}(\lambda, 0)) \exp(-2K_d(\lambda, -0, \theta_0)H), \qquad (3.4)$$

where H is the depth of the water body. Numerical simulations undertaken by Maritorena *et al.* (1994) for realistic hydro-optical conditions indicate that the inaccuracy arising from $k_1, k_2 \equiv K_d$ in eq. (3.4) does not mostly exceed a 15% error in total volume reflectance R_{tot}. The error is largest in a strongly absorbing water body with a single scattering albedo $\omega_0 = 0.2$ and/or in the case of strong hydrosol scattering. A comparison with field measurements suggests that the assumption $k_1, k_2 \equiv K_d$ results on average in an underestimation of $R_{tot}(\lambda, -0)$ throughout the spectrum, if the contrast $(A - R_{wc}) > 0$ (which is the case, for example, of a sandy bottom). Importantly, this underestimation increases with increasing H. A strong underestimation of about 20% at $\lambda < 660\,\mathrm{nm}$ and 30% at $\lambda \approx 680\,\mathrm{nm}$ was equally found when the contrast $(A - R_{wc}) < 0$ (e.g. grass-covered bottom).

In our numerical simulations (Pozdnyakov *et al.*, 2002b) the nadir view angle was assumed and sun zenith angle θ_0 varied from $0°$ to $55°$ with a $10°$ increment. Calculations were conducted for calm surface conditions, and the following vertically homogeneous concentrations of CPAs: $C_{chl}(\mu g/l) = 0.0, 0.5, 1.0, 2.0, 3.0, 4.0, 5.0, 15.0, 20$; $C_{doc}(\mathrm{mg\,C}/l) = 0.0, 0.5, 1.0, 2.0, 5.0, 15.0$; $C_{sm}(\mathrm{mg}/l) = 0.0, 0.5, 1.0, 2.0, 3.0, 4.0, 5.0$; and bottom depths $H(\mathrm{m}) = 1.0, 5.0, 10.0, 20.0, 50.0$. Two types of bottom cover were prescribed: silt and green algae.

The data available on reflectivity spectra of sand, soil humus, vegetation, etc. cannot be directly used for the assessment of albedo, A, because watered surfaces have albedo values essentially lower than dry surfaces with the same cover and they differ in their spectral dependence. Also, soil humus and silts are not identical. Finally, the spectral characteristics of absorption (and, hence, reflection) of solar radiation by benthic vegetation and macrophytes differ from the ones for higher plants (Kondratyev *et al.*, 1999). Fig. 3.21 gives an overview of the spectral variations of coral sand, some types of underwater flora, silica sand, and silt as they were obtained under laboratory conditions with due account taken of the effect of moistening (Estep, 1992).

In our simulations two bottom cover types were explored: silt and the green alga

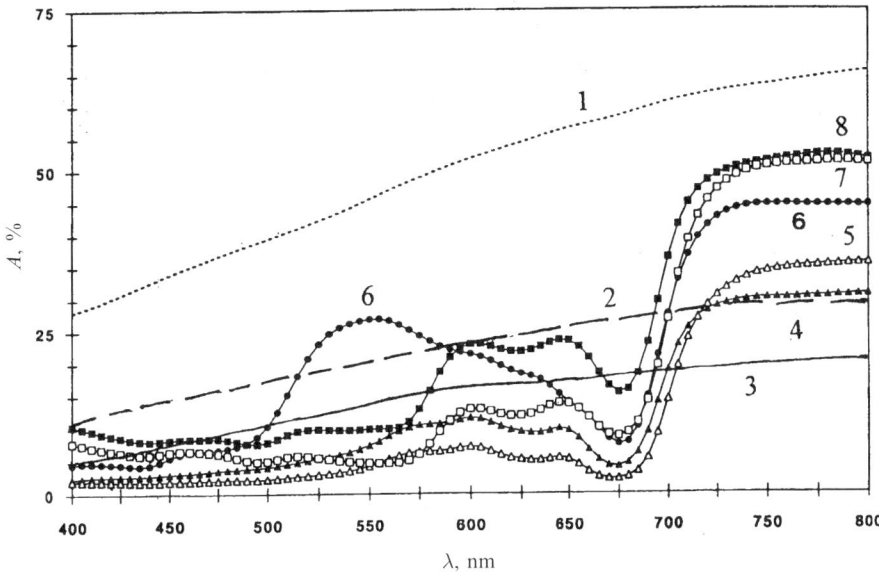

Figure 3.21. Spectral variations of albedo, A, for different bottom cover: 1, coral sand; 2, silicon sand; 3, silt; 4–8, algae (4, *Turbinaria*; 5, *Sargassum*; 6, *Boodlea*; 7, *Porolithon*; 8, *Corallinacea*).

Boodlea (Fig. 3.21, curves 3 and 6, respectively). This choice was prompted by our intention to address such types of bottom cover, which are typical covers in water bodies of temperate latitudes. The trans-spectral effects were included in the simulations as described in Sections 3.1.1–3.1.3. The input values were also specified as in Sections 3.1.1–3.1.3.

A comparison of the results of numerical experiments on the dependence of $\lambda_{dom}(H)$ for a bottom covered with silt or the green alga *Boodlea* (see Table 3.9) with the data for semi-infinite aquatic medium with the same IOPs ($\lambda_{dom}(z = \infty)$) (in both cases the trans-spectral processes are taken into consideration) indicates that the optical impact of the bottom albedo consists in a shift of $\lambda_{dom}(H)$ to longer wavelengths in the visible (Table 3.10).

As should be expected, this shift of the dominant wavelength is more substantial in clear waters and at high sun zenith angles (not shown in Table 3.10). A bottom with green algal cover shifts λ_{dom} more strongly than a bottom consisting of silt; it also increases the value of the shift as compared to a silted bottom.

Enhanced water *turbidity*, due, for example, to an increase in C_{sm} (mg/l) (from 0.1 to 5.0) or C_{chl} (µg/l) (from 0.1 to 20.0) results in a rapid decrease in $\Delta\lambda_{dom}$ even if the bottom depth, H, is only 1 m. This diminution of $\Delta\lambda_{dom}$ becomes particularly pronounced with an increase in the bulk water *absorptivity*, e.g. for C_{doc} (mg C/l) increasing from 0.1 to 20.0. Even at $H = 1$ m, the optical impact of the bottom practically completely disappears (Fig. 3.22).

Table 3.9. Variations in the dominant wavelength λ_{dom} and colour purity p depending on CPA concentrations and bottom depth H(m). Sun zenith angle $\theta_0 = 30°$. The bottom is covered with the alga *Boodlea*.

	$H = 1$		$H = 5$		$H = 10$		$H = 20$	
			$C_{doc} = 0.1\,\text{mg C/l}; \; C_{sm} = 0.1\,\text{mg/l}$					
C_{chl}, µg/l	λ_{dom}, nm	p	λ_{dom}, nm	p	λ_{dom}, nm	p	λ_{dom}, nm	p
0.1	561	0.51	516	0.24	497	0.34	487	0.47
1.0	562	0.53	525	0.26	501	0.27	490	0.35
5.0	564	0.58	548	0.35	518	0.13	507	0.11
10.0	567	0.63	558	0.34	553	0.19	552	0.18
20.0	570	0.68	568	0.33	568	0.31	568	0.31
			$C_{doc} = 0.1\,\text{mg C/l}; \; C_{sm} = 0.5\,\text{mg/l}$					
0.1	562	0.50	512	0.18	496	0.24	492	0.27
1.0	562	0.51	518	0.18	499	0.19	495	0.21
5.0	565	0.56	544	0.22	521	0.10	518	0.10
10.0	567	0.60	558	0.26	555	0.19	555	0.19
20.0	570	0.64	568	0.31	568	0.30	568	0.30
			$C_{doc} = \text{mg C/l}; \; C_{sm} = 1.0\,\text{mg/l}$					
0.1	562	0.48	511	0.14	499	0.16	498	0.17
1.0	563	0.49	517	0.13	503	0.13	501	0.14
5.0	565	0.53	544	0.17	534	0.11	533	0.11
10.0	567	0.56	559	0.23	558	0.20	558	0.20
20.0	570	0.60	569	0.30	569	0.30	569	0.30

As displayed in Tables 3.9 and 3.10, the optical influence of a bottom covered with *Boodlea* on the bulk water colour is no more detectable for $H \geq 20\,\text{m}$. This holds even in very clear/transparent waters.

Our simulations (not exemplified herein) also indicate that the sun zenith angle θ_0 changing from 30° to 55° during daytime summer conditions in the temperate zone controls rather inefficiently the threshold value of CPA concentration vector, above which the bottom optical impact on vanishes. In the case of a silt bottom and highly transparent/clear water, a rather hypothetical case, the bottom optical impact ceases at $H = 10\,\text{m}$ for all sun zenith angles.

For hydro-optical conditions, which are more characteristic of littoral zones of inland water bodies like the large north European Lakes (Ladoga and Onega) or some of the north American (Laurentian) Great Lakes (Ontario, Erie) as well as marine coastal waters, the content of CPAs is often substantial (e.g. in Lake Ladoga, within the littoral zone the concentrations of *chl*, *sm*, and *doc* can be as high as 5–10 µg/l, 2–3 mg/l, and 8–10 mg C/l, respectively (Kondratyev *et al.*, 1999)). Then the optical impact of bottom reflectance on the bulk water colour characteristics is restricted to $H \leq 5\,\text{m}$.

Table 3.10. The dominant wavelength difference $\Delta\lambda_{dom} = \lambda_{dom}(H) - \lambda_{dom}(\infty)$, in nm depending on the concentration vector C (chl, sm, doc) and bottom depth H(m) for sun zenith angle $\theta_0 = 30°$ and bottom cover type (a) *Boodlea* and (b) silt.

Type of bottom cover: *Boodlea*

C_{sm} (mg/l)	$C_{doc}=0.1$ (mg C/l) C_{chl} (μg/l)					$C_{doc}=1.0$ (mg C/l) C_{chl} (μg/l)					$C_{doc}=5.0$ (mg C/l) C_{chl} (μg/l)					$C_{doc}=10.0$ (mg C/l) C_{chl} (μg/l)					$C_{doc}=15.0$ (mg C/l) C_{chl} (μg/l)				
	0.1	1.0	5.0	10.0	20.0	0.1	1.0	5.0	10.0	20.0	0.1	1.0	5.0	10.0	20.0	0.1	1.0	5.0	10.0	20.0	0.1	1.0	5.0	10.0	20.0
										$H = 1.0$															
0.1	76	73	57	15	2	64	61	32	10	2	43	37	21	12	6	43	40	26	18	10	43	40	29	20	11
0.5	70	67	47	12	2	58	53	21	8	2	19	17	11	8	5	21	20	15	11	7	23	22	17	13	7
1.0	64	62	32	9	1	47	40	16	6	2	11	10	7	5	4	12	11	10	8	5	13	13	10	8	5
5.0	8	6	4	2	1	5	4	3	1	0	2	3	2	2	1	2	2	1	1	1	2	1	2	1	0
										$H = 5.0$															
0.1	31	36	41	6	0	42	42	19	3	0	5	3	1	0	0	0	0	0	0	0	0	0	0	0	0
0.5	20	23	26	3	0	28	27	7	1	0	1	1	0	0	0	0	0	0	0	0	0	0	0	0	0
1.0	1	2	1	0	0	0	0	0	0	0	0	0	0	0	0	0	0	0	0	0	0	0	0	0	0
										$H = 10.0$															
0.1	12	12	11	1	0	8	7	3	0	0	0	0	0	0	0	0	0	0	0	0	0	0	0	0	0
0.5	4	4	3	0	0	2	2	0	0	0	0	0	0	0	0	0	0	0	0	0	0	0	0	0	0
1.0	1	2	1	0	0	0	0	0	0	0	0	0	0	0	0	0	0	0	0	0	0	0	0	0	0
										$H = 20.0$															
0.1	2	1	0	0	0	0	0	0	0	0	0	0	0	0	0	0	0	0	0	0	0	0	0	0	0

(*continued*)

Table 3.10. (cont.)

Type of bottom cover: silt

C_{sm} (mg/l)	$C_{doc} = 1.0$ (mg C/l) C_{chl} (μg/l)					$C_{doc} = 1.0$ (mg C/l) C_{chl} (μg/l)					$C_{doc} = 5.0$ (mg C/l) C_{chl} (μg/l)					$C_{doc} = 10.0$ (mg C/l) C_{chl} (μg/l)					$C_{doc} = 15.0$ (mg C/l) C_{chl} (μg/l)				
	0.1	1.0	5.0	10.0	20.0	0.1	1.0	5.0	10.0	20.0	0.1	1.0	5.0	10.0	20.0	0.1	1.0	5.0	10.0	20.0	0.1	1.0	5.0	10.0	20.0
H = 1.0																									
0.1	67	67	57	17	5	62	61	34	13	6	46	40	24	15	8	45	41	28	19	11	43	39	28	19	10
0.5	63	63	47	14	5	57	53	23	11	5	22	20	14	11	7	23	21	16	12	8	22	21	16	12	7
1.0	59	58	32	11	4	47	40	17	9	5	14	13	10	7	6	14	13	11	9	5	12	12	9	7	5
5.0	7	6	4	3	2	5	5	4	2	1	3	3	3	2	1	2	2	2	1	1	1	1	2	1	0
H = 5.0																									
0.1	9	9	24	5	0	17	21	15	2	0	3	2	1	0	0	0	0	0	0	0	0	0	0	0	0
0.5	5	6	14	2	0	10	13	4	1	0	1	1	0	0	0	0	0	0	0	0	0	0	0	0	0
1.0	3	4	6	1	0	6	7	2	0	0	0	0	0	0	0	0	0	0	0	0	0	0	0	0	0
5.0	0	0	0	0	0	0	0	0	0	0	0	0	0	0	0	0	0	0	0	0	0	0	0	0	0
H = 10.0																									
0.1	4	3	5	1	0	3	3	1	0	0	0	0	0	0	0	0	0	0	0	0	0	0	0	0	0
0.5	1	1	1	0	0	1	1	0	0	0	0	0	0	0	0	0	0	0	0	0	0	0	0	0	0
1.0	0	1	1	0	0	0	0	0	0	0	0	0	0	0	0	0	0	0	0	0	0	0	0	0	0

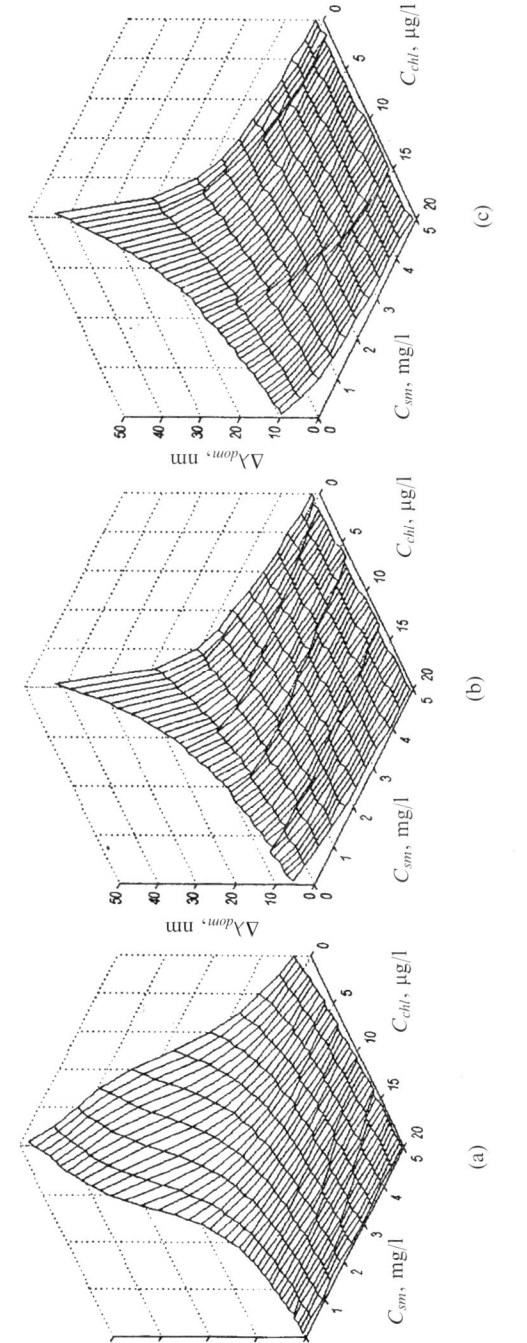

Figure 3.22. Dominant wavelength differences $\Delta\lambda_{dom} = \lambda_{dom}(H) - \lambda_{dom}(\infty)$ in nm for a bottom covered with *Boodlea*. $H = 1$ m, sun zenith angle $\theta_0 = 30°$. C_{doc} (mg C/l) = 0.1 (a); 5.0 (b); 10.0 (c). $\lambda_{dom}(\infty)$ corresponds to the case of semi-infinite aquatic medium.

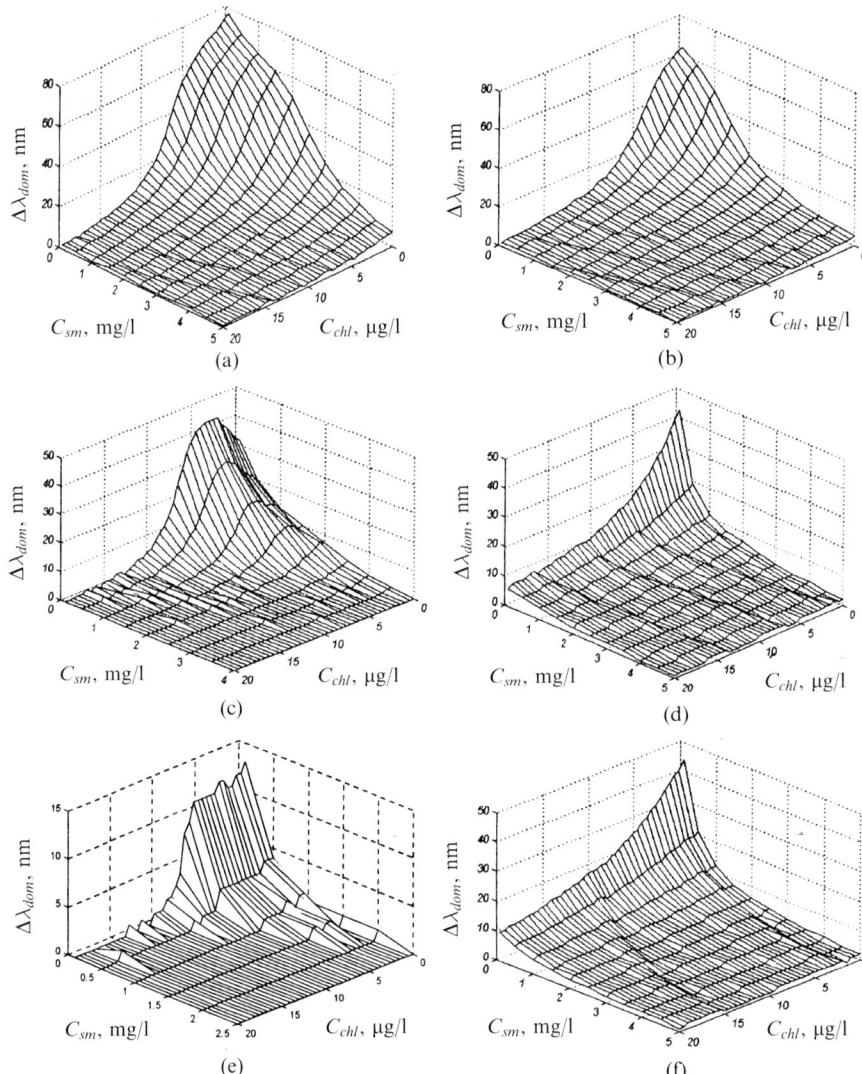

Figure 3.23. Dominant wavelength differences $\Delta\lambda_{dom} = \lambda_{dom}(H) - \lambda_{dom}(\infty)$ in nm for a bottom covered with *Boodlea*. Sun zenith angle $\theta_0 = 30°$. (a) $H = 1\,\text{m}$, $C_{doc}\,(\text{mg C/l}) = 0.1$; (b) $H = 10\,\text{m}$, $C_{doc}\,(\text{mg C/l}) = 0.1$; (c) $H = 1\,\text{m}$, $C_{doc}\,(\text{mg C/l}) = 0.1$; (d) $H = 1\,\text{m}$, $C_{doc}\,(\text{mg C/l}) = 1.0$; (e) $H = 1\,\text{m}$, $C_{doc}\,(\text{mg C/l}) = 5.0$, (f) $H = 1\,\text{m}$, $C_{doc}\,(\text{mg C/l}) = 10.0$. $\lambda_{dom}(\infty)$ corresponds to an optically semi-infinite aquatic medium.

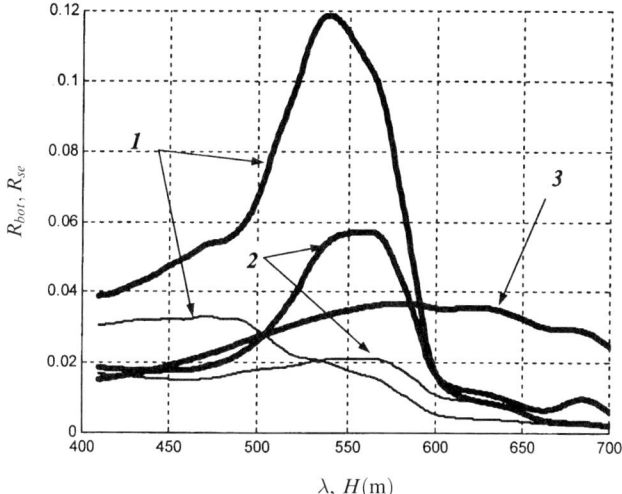

$\lambda,\ H(\text{m})$

Figure 3.24. Total spectral volume reflectance $R_{tot}(\lambda, -0)$ for bottom covered with *Boodlea* (thick line) and volume reflectance for optically semi-infinite aquatic medium $R_{se}(\lambda, -0)$ (thin line). Sun zenith angle $\theta_0 = 30°$. *1*, $H = 5\,\text{m}$, $C_{doc} = 0.1\,\text{mg C/l}$, $C_{chl} = 0.1\,\mu\text{g/l}$, $C_{sm} = 0.1\,\text{mg/l}$; *2*, $H = 5\,\text{m}$, $C_{doc} = 0.1\,\text{mg C/l}$, $C_{chl} = 5.0\,\mu\text{g/l}$, $C_{sm} = 0.1\,\text{mg/l}$; *3*, $H = 5\,\text{m}$, $C_{doc} = 10.0\,\text{mg C/l}$, $C_{chl} = 10.0\,\mu\text{g/l}$, $C_{sm} = 5.0\,\text{mg/l}$.

Analysing Tables 3.9 and 3.10, important additional conclusions can be drawn:

- Suspended minerals are more efficient in neutralizing the optical impact of the bottom than phytoplankton.
- In turn, phytoplankton is more effective than dissolved organics.

It is also noteworthy that the solar zenith angle, θ_0, controls only the shift of the dominant wavelength, but not the bottom depth, H, at which the optical impact can still be traced in $\Delta\lambda_{dom} = \lambda_{dom}(H) - \lambda_{dom}(\infty)$, where $\lambda_{dom}(\infty)$ corresponds to the case of a semi-infinite aquatic medium.

Both for the green alga and silt the maximum value of $\Delta\lambda_{dom}$ occurs for the CPA concentration vector $C(C_{chl}, C_{sm}, C_{doc}) = (0.1, 0.1, 0.1)$ at nearly all simulated bottom depths (Tables 3.9 and 3.10, and Fig. 3.22). The only exception is $C(C_{chl}, C_{sm}, C_{doc}) = (5.0, 0.1, 0.1)$ for $H = 5\,\text{m}$ (Fig. 3.23(c)). This particularity originates from an additional strong reflection at about 565 nm. Its height equals or even slightly exceeds the intensity of the reflection band at 530 nm (Fig. 3.24).

Figure 3.25 illustrates the threshold values of the CPA concentration vector $C(C_{chl}, C_{sm}, C_{doc})$, at which the optical influence of the bottom covered with the green alga *Boodlea* is still affecting λ_{dom}. Analogous results (not shown here) were obtained for a silted bottom. Obviously, as the bottom depth, H, increases, the threshold values of the concentration vector $C(C_{chl}, C_{sm}, C_{doc})$ decrease steadily: at $H = 20\,\text{m}$, the optical impact of the 'green' bottom has nearly vanished at $C(0.1, 0.1, 0.1)$, whereas in the case of a silted bottom its optical influence (at the same C) ceases already at $H = 10\,\text{m}$.

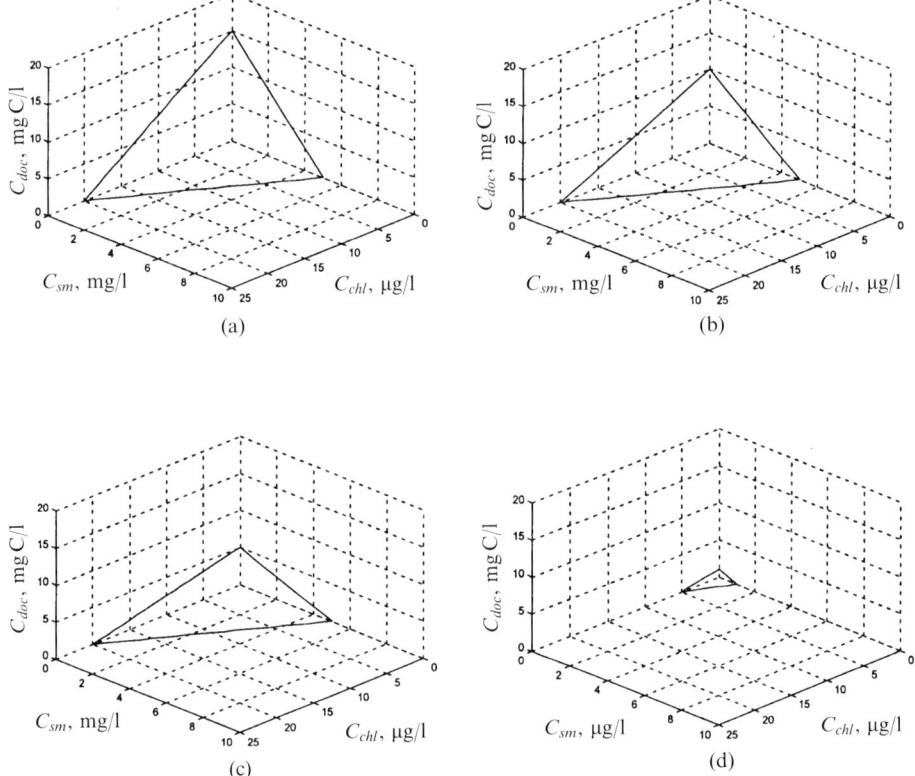

Figure 3.25. *chl*, *sm*, and *doc* concentrations at which the optical impact of bottom albedo (for *Boodlea*) on water colour becomes negligible for a sun zenith angle $\theta = 30°$ and several depths: (a) 1 m, (b) 5 m, (c) 10.0 m, and (d) 20.0 m.

Thus, in optically shallow waters with the two bottom cover types considered the simulated optical impact on the bulk water colour results in:

(a) A more rapid displacement of λ_{dom} to its *end-point* (asymptotic) value as the concentration vector $C(C_{chl}, C_{sm}, C_{doc})$ increases compared to the case of optically deep waters, trans-spectral processes included.
(b) The *end-point* or asymptotic value of λ_{dom} in optically shallow waters is determined not solely by the concentration vector $C(C_{chl}, C_{sm}, C_{doc})$ (as it takes place in optically deep waters), but also by the bottom depth H.
(c) As the bottom depth, H, increases also the differences both in the rates of λ_{dom} displacement to the asymptotic (*end-point*) regime and the absolute values of end-point λ_{dom} decrease; at $H \geq 10$–20 m (depending on bottom cover) all differences disappear.
(d) Colour purity, p, in optically shallow waters does not follow the characteristic features/regularities established for optically deep waters; within the range of concentrations exploited in this study, the value of p can either increase or drop

depending on the concrete combinations of H and $C(C_{chl}, C_{sm}, C_{doc})$; when the optical impact of bottom is strong, p can increase up to 0.5 or even 0.8; however, with decreasing optical influence of the bottom, the colour purity p attains values characteristic of deep waters with the respective concentration vector C.

These computational results are in general agreement with field data from shallow waters (Pomeranian Bight in the vicinity of the Odra Bank) in the Baltic Sea (Ohde and Siegel, 2001). The measurements were conducted over a silted bottom and were supplemented by CZCS colour images. Using a logarithmic version of eq. (3.4),

$$\ln\left[\frac{A(\lambda) - R_\infty(\lambda, -0)}{R_{tot}(\lambda, -0) - R_\infty(\lambda, -0)}\right] = 2K_d H, \qquad (3.5)$$

Ohde and Siegel (2001) obtained a regression relationship at $\lambda = 520\,\text{nm}$ between the left side of eq. (3.5) and H. The linear relationship was valid only up to $H = 13\text{--}15\,\text{m}$ and average $C_{chl} = 3\text{--}5\,\mu\text{g/l}$. Although the complete concentration vector $C(C_{chl}, C_{sm}, C_{doc})$ is not provided for the study area, the results of our computations of a limit value of the bottom depth, H, at which the optical influence of the bottom vanishes compares well with the results reported by Ohde and Siegel (2001).

4

Quantitative interpretation of satellite colour images over non-Case I waters: numerical simulations of the inverse problem

4.1 CONCEPTUAL APPROACH TO THE SOLUTION OF THE INVERSE PROBLEM

As seen in the preceding chapter, water colour results, in addition to absorption and elastic scattering, also from water Raman (inelastic) scattering, fluorescence by phytoplankton and dissolved organic matter, and, if waters are optically shallow, from bottom reflectance as well (Fig. 4.1).

Importantly, the contribution of all these processes of photon interactions with the aquatic medium to water colour is highly variable and depends on a wealth of factors such as the concentration vector $C(C_{chl}, C_{sm}, C_{doc}, \ldots)$, *geometry* of the air–water interface, *intensity* of illumination by direct sun and diffuse sky radiation, bottom cover type (when applicable), and varying spectral cross-sections of colour producing agents (CPAs), etc.

This implies that water colour per se and its proxies, spectral volume reflectance $R(\lambda, -0)$, upwelling spectral radiance $L_u(\lambda, +0)$, and remote sensing reflectance $R_{rs}(\lambda, +0)$ are generally *nonlinearly* related to water quality parameters. Therefore, when solving the inverse problem, it is mandatory to apply an algorithm, which takes into account the *compound* nature of the parameter (i.e. water colour or its proxies) to be used in the retrieval procedure.

It also implicitly indicates that as far as all CPAs affect water colour and its proxies, irrespective of optical simplicity of natural waters, the retrieval procedure should provide for a simultaneous retrieval of the major CPAs (and not only *chl*, even if it is the sole incentive for remote sensing, as for Case I waters).

Categorizing the retrieval approaches used so far, it is possible to differentiate between them following Cracknell *et al.* (2001). Historically, it was first the *empirical* (statistical) approach. Based on the establishment of statistical relations between remotely measured values and some water quality parameter(s) determined *in situ*, this approach suffers from ignoring causal relationships between the magnitudes

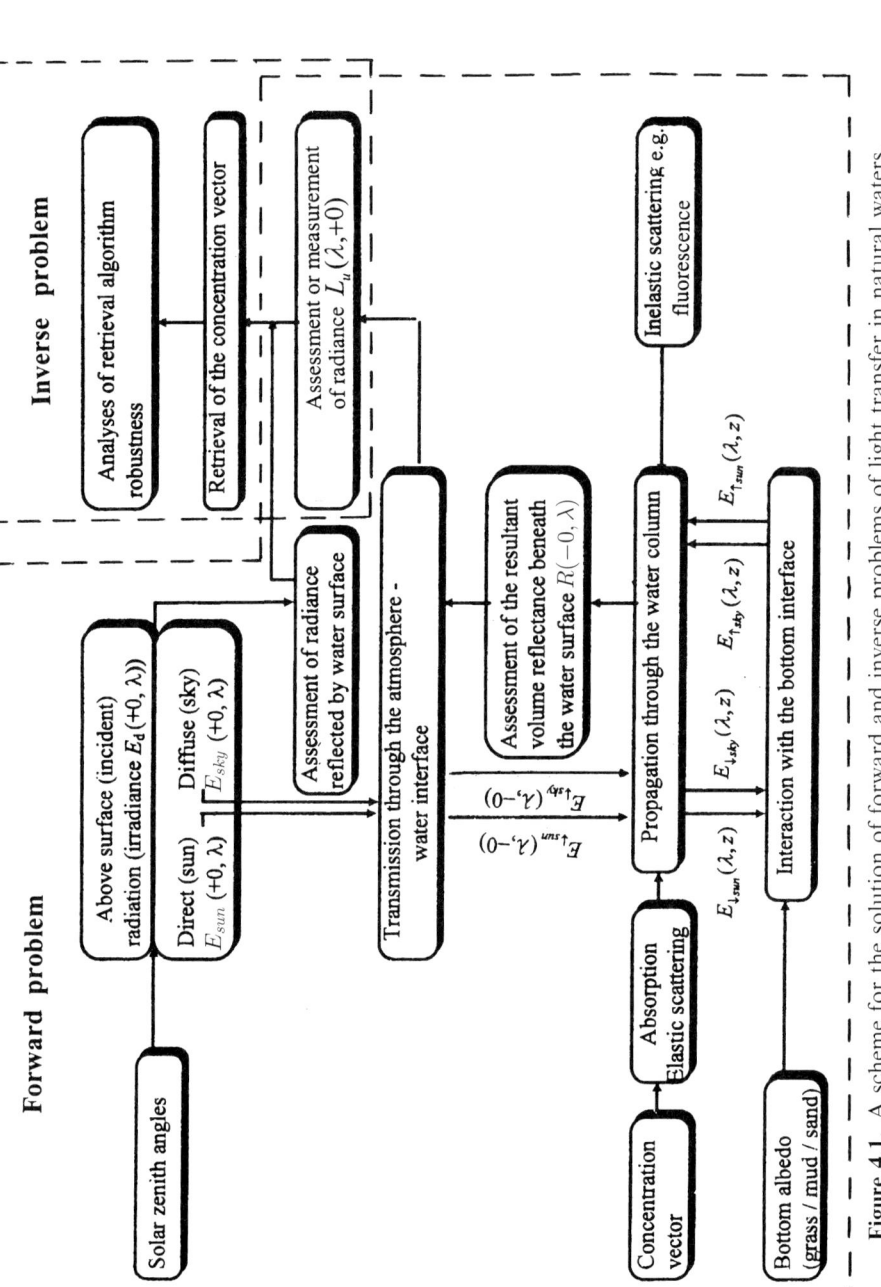

Figure 4.1. A scheme for the solution of forward and inverse problems of light transfer in natural waters.

involved. It is applicable only to a concrete set of conditions under which it was developed, and thus is not suitable for other locations/datasets. The *semi-analytical* approach uses spectra from characteristic regions of the image and knowledge of the spectral characteristics of CPAs. It relates spectral *band-ratios* to the concentrations of one or more water quality parameter. Finally, the *analytical approach* relates the optical properties of the water column to water colour or its proxies via a physical process.

It is clear, that within the framework of at least the analytical approach, which employs water colour proxies to infer CPA concentrations, a certain *iterative* procedure should be used, because the existing relations (see, for example, eqs. (1.18–1.20), (1.24), (1.25)) between $R(\lambda, -0)$ or $R_{rs}(\lambda, +0)$ and the IOPs of water do not include trans-spectral effects and the impact of bottom reflection. To also account for these processes, one of the possible ways consists in: firstly, exploiting a retrieval procedure using one of the above equations; secondly, after having deter-mined a first guess concentration vector, calculate the impacts of trans-spectral and bottom reflection effects; and, thirdly, subtract their signatures from the measured spectrum of the water colour proxy. Finally, apply again the retrieval procedure to obtain more accurate values of the concentration vector and, if converging, repeat this procedure.

The above scheme leaves out of consideration the issue of vertical distributions of CPAs within the water column. Kondratyev *et al.* (1990) analysed through Monte Carlo simulations the impact of vertical inhomogeneity of water optical properties on the light emerging from beneath the water surface. Their calculations indicate that in turbid waters (Lake Ladoga coastal waters were taken as an example) this effect generally does not account for more than ~5% of the legitimate signal vari-ability, and it is even less when the maximum in the vertical distribution of CPAs is located lower than the middle of the local euphotic depth. However, for clearer waters the vertical inhomogeneity of water optical properties may play a more important role (see, for example, Frette *et al.*, 2001) and hence deserves a closer consideration.

4.2   ALGORITHMS FOR THE RETRIEVAL OF WATER QUALITY PARAMETERS IN NATURAL WATERS: A BRIEF HISTORICAL OVERVIEW

4.2.1   Case I waters

As emphasized in the Preface, for pristine/offshore oceanic waters qualifying as Case I waters (recall, these are waters where phytoplankton, together with accompanying and co-varying products of their life cycles, as well as some microscopic organisms such as flagellates, bacteria and viruses, are the principal agents determining the variations in optical properties of the water), it was believed for a long time that considerable simplifications are possible. Through a combination of statistical and semi-analytical universally applicable relationships between either the upwelling radiance $L_u(\lambda, +0)$ or remote sensing reflectance $R_{rs}(\lambda, +0)$ ratios at two or more wavelengths and the desired parameter, normally just the concentration of chlorophyll-a, a simple and robust retrieval scheme had been developed.

The preferred choice was the ratio between two spectral bands, often the water leaving radiance $L_u(\lambda_i, +0)$ in the blue close to the wavelength of maximum absorption by chlorophyll (λ_i) normalized by $L_u(\lambda_j, +0)$ at a wavelength λ_j, for which $L_u(\lambda, +0)$ could be assumed quasi-independent of chlorophyll-a (leaving alone water per se). This leads to very simple algorithms of the form:

$$C_{chl} = A[L_u(\lambda_i, +0)/L_u(\lambda_j, +0)]^{-B}, \tag{4.1}$$

where A and B are regression coefficients.

When doc and sm are present at low concentrations the retrieval of chlorophyll content results improves when λ_i and λ_j are shifted to longer wavelengths. Therefore, various λ_i/λ_j pairs (443/520 nm; 443/550 nm; 520/550 nm; 520/670 nm) have been tried for a given combination of C_{chl}, C_{doc}, and C_{sm} to best fit eq. (4.1) (NASA, 1993; Siegel *et al.*, 1999; for other references see Kondratyev *et al.*, 1999). Naturally, the regression coefficients A and B in eq. (4.1) are area- and season-specific, since the optical properties of CPAs constituting the concentration vector $C(C_{chl}, C_{sm}, C_{doc})$ vary with space and time.

To avoid this shortcoming that obviously limits the operational use of band-ratio regression algorithms, spectral derivatives were also suggested to improve the retrievals of CPAs in Case I waters. Grew (1981) found that the ratio

$$G_{mn}(\lambda_i) = \frac{L(\lambda_i)^2}{L(\lambda_m)\, L(\lambda_n)},$$

where $i = 1, 2, 3, \ldots, k$ represents the centre spectral channel or band number, and m, n stand for adjacent ones, varies significantly less with solar elevation, sea state, and cloud cover than the simple radiance ratios. Based on this premise and adhering to the two-band (blue–green) ratio methodology, Campbell and Esaias (1983) suggested the following chlorophyll retrieval algorithm:

$$\lg C_{chl} = a - b \lg \frac{L(490)^2}{L(460)\, L(520)}, \tag{4.2}$$

with $a = 1.43, b = 10.02$. Allegedly, this algorithm enhances spectral features of the water medium while effectively eliminating variations due to changes in incident irradiance. Since the irradiance upwelling from pure water displays a strong negative curvature at 490 nm, an increase of chlorophyll-bearing algae tends to decrease monotonically this negative curvature. Hoge and Swift (1986) analysed the suitability of existing or proposed ocean colour sensor bands (CZCS, OCM(ERS-1), OCI(NOAA-K), OCI(SPOT-3)) and again suggested the use of spectral bands in the blue–green portion of the spectrum. Their approach was further extended by Hoge and Lyon (1996). However, when Lee and Carder (2000) recently compared the operational retrieval efficiency of band-ratio and spectral curvature algorithms, they claimed that band-ratio algorithms still have greater potentials to be applied to a wider range of oceanic waters, whereas spectral-curvature algorithms show stable performance only when the data set falls into an appropriate narrower range. Hence, the spectral derivative method seems applicable as a synoptic colour classification method of natural waters (Lahet *et al.*, 2001a) only for well-defined areas and water bodies.

Meanwhile, further work aiming at improving band-ratio algorithms has been carried out. At the SeaWiFS Bio-optical Algorithm Mini-workshop (SeaBAM) in January 1997 a revision of operational band-ratio algorithms was suggested. While the ocean chlorophyll algorithm 2 (OC2), a modified cubic polynomial function that uses the ratio $R_{rs}(490, +0)/R_{rs}(555, +0)$ and simulates well the sigmoidal pattern revealed between log-transformed radiance ratios and *chl*, was chosen as the pre-launch SeaWiFS operational *chl-a* algorithm (for OC2, OC2v2, OC2v3 parameters see, for example, Smyth *et al.*, 2002), an improved performance was obtained using the ocean chlorophyll algorithm 4 (OC4). This is a four-band (443, 490, 510, 555 nm) maximum band ratio formulation (O'Reilly *et al.*, 1998):

$$C_{chl} = 10^{(a_0 + a_1 \tilde{R}_{rs} + a_2 \tilde{R}_{rs}^2 + a_3 \tilde{R}_{rs}^3)} + a_4, \tag{4.3}$$

where R_{rs} is the greatest log-ratio among $(R_{rs}(443)/R_{rs}(555), R_{rs}(490)/R_{rs}(555),$ and $R_{rs}(510)/R_{rs}(555)$, and $a_0 = 0.4708; a_1 = -3.8469; a_2 = 4.5338; a_3 = -2.4434; a_4 = -0.0414)$.

As can be seen, neither trans-spectral processes nor non-co-varying CPAs (i.e. *sm* and *doc*) have been taken into account for all these algorithms (Gordon and Morel, 1983; Morel, 1988; Loisel and Morel, 1998; O'Reilly *et al.*, 1998; Morel and Maritorena, 2001).

However, with the increasing accuracy of satellite sensors, and more demanding 'customer' requirements with respect to quality/precision of remote sensing retrievals (e.g. Durand *et al.*, 1998; Cracknell *et al.*, 2001), the Case I water community is at present searching for new algorithms capable of meeting the above challenges. To achieve these goals, many workers resort to area-specific algorithms. For instance, Gohin *et al.* (2002) suggested using the OC4 algorithm as an input parameter for a regression area-specific algorithm OC5 of the form $OC5 = OC4 - A_1(OC4 - 0.55)A_2$, where A_1 and A_2 are estimated at 0.18 and 2.0, respectively.

Some studies have addressed the possible inaccuracies inherent in the design of space sensors, e.g. SeaWiFS, that are due to the width of spectral bands (Wang *et al.*, 2001), and digitization errors (Hu *et al.*, 2000a), instability of calibration in several channels as well as imperfection of atmospheric and water surface corrections (Gould *et al.*, 2001).

Nevertheless, the major attention is still paid to the methodological aspects of CPA retrievals. Alongside purely empirical/statistical models, a number of semi-analytical models have been suggested. In an attempt to decouple the phytoplankton chlorophyll concentration from the concentrations of other CPAs co-existing in Case I waters, Garver and Siegel (1997) developed a model, in which the combined optical impact of *doc* and detritus was assumed *independent* of C_{chl}. *Cross-sections* of the chosen CPAs were used, and a *nonlinear* least-squares inversion procedure was employed. In addition to an area-specific band-ratio retrieval algorithm, an inverse modelling technique was also applied by Siegel *et al.* (1999) separating different water constituents in open (and coastal) waters of the Baltic Sea.

Based both on (a) evidence (Siegel and Michaels, 1996; see also Karhu and Mitchel, 2001) that in offshore waters (i.e., generally, Case I waters) there might be areas with high dissolved organics not co-varying with chlorophyll concentrations, and (b) the dependence of $a_{chl}^*(\lambda)$ on the pigment packaging effect, Carder *et al.* (1999) have suggested for MODIS, or sensors with similar spectral channels, semi-analytical algorithms for a *simultaneous* retrieval of *chl* as well as phytoplankton and *doc* absorption, depending on bio-optical domains defined according to their nitrate-depletion temperatures.

Based on a developed underwater optics model, Ammenberg *et al.* (2002) suggested a set of semi-analytical algorithms, relating through regression the ratio of *remote sensing reflectance* $R_{rs}(\lambda)$ at 705 and 664 nm (for *chl*), 664 and 550 nm (for the absorption coefficient of coloured dissolved organic matter, a_{CDOM} at $\lambda = 420$ nm), as well as an algorithm relating $R_{rs}(705)$ to *sm*.

However, many semi-analytical models suggested more than ten years ago (e.g. Sathyendranath *et al.*, 1989), and even those which have been developed only recently (e.g. Reynolds *et al.*, 2001) remain within the paradigm of the 'classical' definition of Case I waters. The main task proposed by them is to increase the statistical data in order to attain more accurate and more universally applicable relationships between optical properties of CPAs and the concentration of *chl* (e.g. Loisel and Morel, 1998; Morel and Maritorena, 2001).

When reappraising the bio-optical properties of oceanic waters, and suggesting some modified and more accurate parameterizations relating absorption and back-scattering coefficients for particulate suspended matter, Morel and Maritorena (2001) had to acknowledge that for reasons as yet unknown the parameterizations derived for the Pacific Ocean's 'classical' Case I waters, prove to be inadequate, for instance, for the Mediterranean Sea. Dierssen and Smith (2000) report that in Antarctic offshore waters the *chl* retrieval algorithms recommended by SeaBAM for universal use in Case I waters underestimate concentrations of *chl* by roughly a factor of 2. Karhu and Mitchel (2001) faced the necessity of refusing OC2 and OC4 SeaBAM algorithms in favour of their own local algorithms to attain more acceptable accuracies in *chl* retrievals. These and some other problems associated with more accurate remote sensing of C_{chl} in Case I waters, prompted extended deliberations on the part of the above authors as to the desirability of arriving eventually at an *analytical* ocean reflectance model. Within the framework of such a model both bulk water coefficients a and b_b (or better the volume scattering function) have to be reconstructed as a sum of *separate* and possibly *independent* contributions by all significant CPAs.

Trans-spectral processes were also reconsidered as, probably, important players in the formation of colour or its proxies in Case I waters. Recently Gordon (1999) reexamined the contribution of water Raman scattering and came to the conclusion that in waters with $C_{chl} \leq 1$ µg/l the Raman fraction in upwelling radiance exceeds

8% at the wavelengths of interest for ocean colour remote sensing and therefore cannot be ignored in processing the SeaWiFS images. Based on the earlier model for reflectance due to Raman scattering Sathyendranath and Platt (1998) and Sathyendranath et al. (2001) investigated an analytical model of Case I water reflectance that accounts for the water Raman scattering contribution to total volume reflectance.

The fluorescence by chlorophyll has been considered by some workers as an important process in determining the spectral distribution of volume reflectance and upwelling radiance in the red portion of the visible spectrum. The fluorescence peak of chlorophyll at 685 nm was suggested as a means for remotely retrieving the content of phytoplankton or rather chlorophyllous pigments as their proxy in Case I waters (Neville and Gower, 1977; Doerffer, 1981; Gower and Borstad, 1990, 1993; Gower et al., 1999; and some others, for references, see also Kondratyev et al., 1999). The Medium Resolution Imaging Spectrometer (MERIS), which was launched by the European Space Agency on ENVISAT on 28 February 2002, has a band at 681.25 nm for *chl* fluorescence (rather than at 685 nm because of the presence of atmospheric O_2 absorption at $\lambda \geq 687$ nm). Gower et al. (1999) suggest using five bands in the spectral range 620–754 nm for inferring C_{chl} and interpreting spectral features of $L_u(\lambda, +0)$. The algorithm proposed for estimating the *chl* fluorescence signal, called fluorescence line height (*FLH*), involves subtraction of an interpolated baseline designed to represent the spectrum of $L_u(\lambda, +0)$ in the absence of fluorescence. In the simplest case this interpolation is linear, and the *FLH* retrieval algorithm may be written in the form:

$$FLH = L_u(\lambda_2, +0) - L_u(\lambda_1, +0) - (L_u(\lambda_3, +0) - L_u(\lambda_1, +0))\frac{(\lambda_2 - \lambda_1)}{(\lambda_3 - \lambda_1)}, \qquad (4.4)$$

where λ_2 is the MERIS band centred on the *chl* fluorescence, while bands λ_1 and λ_3 determine the baseline. Numerical and field experiments indicated that the optimal combination of λ_2, λ_1, and λ_3 is 682 nm, 665 nm, and 705 nm, respectively.

The relationship between *FLH* and C_{chl} obtained by Gower et al. (1999) on the coast of British Columbia and in the Baltic Sea appear to be linear but very different in slope depending upon the location. Generally, this linearity only holds around $C_{chl} = 1$ µg/l and strongly departs at C_{chl} close to 0.1 µg/l, and $C_{chl} > 10$ µg/l (at $C_{chl} \geq 30$ µg/l the slope even changes sign).

Extending the *FLH* algorithm to waters containing an enhanced amount of suspended matter, Gower et al. (1999) found that although suspended matter strongly reduces the *chl* fluorescence peak, in full qualitative agreement with our results of numerical modelling (see Chapter 3), the value of *FLH* remains, however, approximately stable.

The issue of *retrieval accuracy* remains open under a significantly reduced *absolute* value of the *chl* fluorescence peak in the case of enhanced suspended and dissolved organic matter. In addition, as will be emphasized below, the absorption by water per se becomes strong in the red portion of the spectrum (see Table 2.1). In situations with a vertically inhomogeneous *chl* distribution, e.g. submerged maximum, the *chl* fluorescence peak will be drastically reduced. Assessed by remote sensing via *chl* fluorescence alone, such waters will appear less productive than in reality.

Returning to the band-ratio approach, the increasing awareness of lack of co-variance between the concentration of chlorophyll and the content of other CPAs in Case I waters stimulated a wealth of studies exploiting the neural networks approach (NN). Their major asset is the ability to deal with a wide range of nonlinear continuous functions. This property and the ability of NN to detect noise can be exploited, in certain conditions, to filter the noise during model training, which is crucially important when processing real data.

Heinemann and Fischer (1997) and Schiller and Doerffer (1999), together with Keiner and Brown (1999) and Pozdnyakov and Lyaskovsky (1998), pioneered the application of NN to retrieve natural water colour variables. Since then the number of publications using NN to infer water quality parameters in oceanic waters has considerably increased.

There are many different types of neural networks but one of the most commonly used in remote sensing is the multilayer perceptron (MLP) (Atkinson and Tatnall, 1997): the MLP generally consists of three layers. The input layer neurons are the elements of a feature vector which might consist of radiances at certain wavelengths. The second layer is the internal or 'hidden' layer. In the third layer, the number of neurons equals the number of parameters to be determined. Each neuron in the network is connected to all neurons in both the preceding and subsequent layers by connections with associated weights.

The input signals are transferred to the neurons in the next layer in a feed-forward manner. As the signal propagates from neuron to neuron, it is modified by the appropriate connection weight. The receiving neuron sums the weighted signals from all neurons to which it is connected in the previous layer. The total input that the jth neuron receives is weighted in the following way:

$$net_j = \sum_{i=1}^{N} \omega_{ji} o_i, \qquad (4.5)$$

where ω_{ji} is the weight of relationship between neuron i and neuron j, and o_i is the output from neuron i. The output from a given neuron j is then obtained from:

$$o_j = f(net_j). \qquad (4.6)$$

The function f is usually a nonlinear sigmoid function. It is applied to the weighted sum of inputs before the signal reaches the next layer. When the signal reaches the output layer, the network output is produced. The created network should be first trained so that it could generalize and predict outputs from inputs that it has not processed before. A training pattern is fed into the neural network and the signals are forwarded. After that, the network output is compared to the true output, the error is then computed and back-propagated through the network. As a result, the connection weights are modified following the generalized rule:

$$\Delta\omega_{ji}(n+1) = \eta(\delta_j o_i) + \alpha\Delta\omega_{ji}(n), \qquad (4.7)$$

where η is the learning rate parameter, δ_j is an index of the error change rate, and α is the momentum parameter. The training is conducted until the output error reaches a desired level of accuracy. The trained neural network is then tested against some verification data to assess the network performance.

Gross *et al.* (1999) used NN for retrieving *chl* in Case I waters, in order to examine the differences between the NN (a multilayered perceptron) and the cubic polynomial that is usually applied to retrieve C_{chl} (see above) as well as for assessing the capability of NN to filter noise (e.g. radiometric noise and/or the residual atmospheric correction errors). The best NN architecture was determined using the constructing methodology residing in intensive application of cross-validation (Bishop, 1995). Exploiting data sets simulated on the basis of the Morel bio-optical model (Morel, 1988), Gross *et al.* (1999, 2000) have given solid evidence that NNs are able to perform more efficiently than the cubic-polynomial regression algorithms, and are considerably more stable in dealing with noisy input data.

Earlier, in order to reduce the number of input neurons and to increase the information content of the input, Heinemann and Fischer (1997) suggested using $R_{rs}(\lambda, +0)$ values after a principal component analysis. Buckton *et al.* (1999) trained an algorithm using the Levenberg–Marquardt method (for references see Kondratyev *et al.*, 1999), which converges faster to global minima. Also the singular value decomposition (SVD) technique has been used as pre-processing for data reduction. The NN algorithm was able to restore the concentrations of three CPAs, namely *chl*, *sm*, and *doc*.

Working on classification of space images, Warrender and Augusteijn (1999) suggested fusing several classification methods using Bayesian techniques with Markov random fields added. NN was one of the classification methods involved and the back-propagation scheme was used. It is characteristic of this scheme that the learning method is slow and convergence to a solution is not guaranteed: the network may get trapped in a *local* (not *global*) minimum. Correct execution depends on the optimal choice of the number of neurons in the hidden layer(s). Although this choice is still an art rather than a science, a cascade-correlation architecture approach suggested by Fahlman and Lebiere (1990) can seriously improve NN creation: owing to this approach the network builds its internal structure incrementally during training. Application of the Fahlman and Lebiere method proved to be efficient in image classification as well as in other remote sensing applications (also see Augusteijn *et al.*, 1995). Also for classification purposes, Kavzoglu and Mather (1999) suggested applying a procedure of *pruning* (magnitude-based *pruning*, optimum brain *damage*, and optimal brain *surgery*) to artificial neural networks. By *pruning*, the network size can be significantly reduced without reducing the accuracy of the classification results. Supposedly, this modification/improvement of NN is also applicable to NN designed for inferring water quality parameters from space images.

Neural network techniques are also increasingly applied to spectral radiative-transfer modelling, in order to considerably reduce computation time. Schwander *et al.* (2001) used this technique for restoring a nearly continuous spectrum (153 wavelengths) of incident global radiation in the range from 280 to 700 nm from input consisting of only seven wavelengths together with solar zenith angle and total ozone amount. This numerical experiment might be important for improved atmospheric correction and in-water photon propagation codes.

Jointly with seeking solutions of inverse problems from the perspective of inferring water quality parameters, considerable attention was and is paid not only to retrieving bulk optical properties of waters from remotely sensed water colour (or rather its proxies, $R(\lambda, -0)$, $R_{rs}(\lambda, +0)$, and $L_u(\lambda, +0)$) but also to reconstruct water colour proxies from *in situ* measurements of total absorption and scattering of the aquatic medium. However, the limited scope of this book does not allow us to accommodate this special area of research. Many new publications (e.g. Roesler *et al.*, 1989; Leathers *et al.*, 1999; Barnard *et al.*, 1999; Pinkerton *et al.*, 1999; Boynton and Gordon, 2000; Loisel and Stramski, 2000; Lee *et al.*, 2001; Loisel *et al.*, 2001; Aguirre-Gomez *et al.*, 2001; Lahet *et al.*, 2001b) have recently appeared. However, we will concentrate on those aspects of aquatic optics that are directly applicable to the interpretation of radiation upwelling from Case II waters.

4.2.2   Case II waters

Historically, the *empirical/statistical* approach to infer *chl* (see Section 4.2.1) was extended to Case II waters. NASA workers (McClain and Yen, 1994) suggested a supervised band-ratio algorithm, which was simultaneously applicable to Case I and Case II waters:

$$C_{chl} = 3.33 \left[\frac{L_u(520)}{L_u(550)} \right]^{-2.40}. \tag{4.8}$$

A constraint exists for eq. (4.8): it should only be used provided that C_{chl}, as calculated through eq. (4.9),

$$C_{chl} = 1.13 \left[\frac{L_u(443)}{L_u(550)} \right]^{-1.71} \tag{4.9}$$

falls into the range $C_{chl} \leq 1.5 \, \mu\text{g/l}$. The retrieval algorithm (4.8) can still be used when $C_{chl(4.8)} > 1.5 \, \mu\text{g/l}$, if C_{chl}, as calculated through eq. (4.9), is less than $1.5 \, \mu\text{g/l}$; otherwise, eq. (4.9) is to be used.

This complication of the retrieval procedure could be circumvented, according to McClain and Yen (1994), via a 'three-channel' algorithm:

$$C_{chl} = 5.56 \left[\frac{L_u(443) + L_u(520)}{L_u(550)} \right]^{-2.252}. \tag{4.10}$$

However, Gordon and Wang (1994a) insisted that the $L_u(443)/L_u(550)$-based algorithm was more efficient when used in the form of a thirdorder polynomial in r_L:

$$\log 3.33 C_{chl} = -1.2 \log r_L + 0.5(\log r_L)^2 - 2.8(\log r_L)^3, \tag{4.11}$$

where

$$r_L = \frac{1}{2} \frac{[L_u(443)]_N}{[L_u(550)]_N},$$

and $[L_u(\lambda, +0)]_N$ is the normalized spectral water-leaving radiance calculated via

$$tL_u(\lambda) = [L_u(\lambda)]_N \cos\theta_0 \exp\left[-\left(\frac{\tau_r}{2} + \tau_{oz}\right)\left(\frac{1}{\cos\theta_0} + \frac{1}{\cos\theta}\right)\right], \qquad (4.12)$$

where t is the diffuse transmittance of the atmosphere, τ_r is the Rayleigh optical thickness (molecular scattering), τ_{oz} is the optical thickness of ozone, and θ_0 and θ are, respectively, the sun zenith angle and angle of viewing of a flat water surface.

A very similar algorithm has been suggested by Sturm (1993) which was called the 'European method'. Based on the Case I water model developed by Morel (1988), a regression has been established between $C_{chl}(\mu g/m^3)$ and the radiance ratio of the first and third spectral CZCS channel as well as the second and third:

$$\ln C_{chl} = 0.768 - 2.61 \ln X + 0.791(\ln X)^2 - 0.388(\ln X)^3; \qquad (4.13)$$

$$\ln C_{chl} = 1.395 - 8.739 \ln Y + 5.7286(\ln Y)^2 - 27.9504(\ln Y)^3, \qquad (4.14)$$

with $X = R(443)/R(550)$, $Y = R(520)/R(550)$, whereby $R(\lambda)$ is the subsurface volume reflectance.

Algorithms (4.13) and (4.14) were used for retrievals of C_{chl} in Case I waters with moderate and high phytoplankton chlorophyll concentrations, respectively. The value of $R(443)$ was used as a criterion for the transition from algorithm (4.13) to algorithm (4.14): if the value of $R(443) \leq 0.4$, then eq. (4.14) should be used.

In practice the threshold corresponds to $C_{chl} = 2$–$3\ \mu g/l$. As another criterion for automated differentiation between Case I and Case II waters, $R_{lim}(550)$ is applied:

$$R_{lim} = \exp[1.05 - 0.02\ln X - 0.429(\ln X)^2 + 0.094(\ln X)^3]. \qquad (4.15)$$

Case II waters could also be identified, provided $C_{chl} \leq 10\ \mu g/l$, using (4.14).

According to Sturm (1993), C_{chl} in Case II waters can also be determined via polynomial algorithms based on the ratios $X = R(443)/R(550)$, $Y = R(520)/R(550)$. However, in the light of the considerations presented above, when discussing the spectral variations of $R(\lambda, -0)$ and the displacements of the 'hinge-point' driven by increasing concentrations of CPAs a satisfying performance seems unlikely.

The complexity of the composition of Case II waters and their associated optical properties prompted a further development of water quality retrieval techniques.

Still adhering to regression algorithms, but already applying a hydro-optical water model based on specific inherent hydro-optical coefficients of phytoplankton and suspended minerals, Tassan (1994) suggested a system of algorithms for CPA retrieval. The algorithms are based upon established regression relationships between major CPAs for Case II waters (what seems surprising in view of the complexity of Case II water). For the Gulf of Naples and the northern margins of the Adriatic Sea the following interdependences between chl, sm, and doc were established:

$$\log C_{sm} = -A_1 + B_1 \log C_{chl}, \qquad (4.16)$$

$$\log a_{doc}(440) = -A_2 + B_2 \log C_{chl}, \qquad (4.17)$$

where A_1, A_2, B_1, B_2 are regression coefficients with values 0.25, 0.026, 1.20, 1.28, and

0.57, 0.59, 0.47, 0.38 for the waters of the Gulf of Naples and the northern margins of the Adriatic Sea, respectively. For $a_{doc}(440)$, relation (2.3) has been adopted with $s = -0.014 \, \text{nm}^{-1}$.

The regression algorithms proposed by Tassan (1994) are the following:

$$\log C_{chl} = 0.0664 + 0.0462 \log(X_{chl}) - 4.144 \log(X_{chl})^2 \quad (0.025 \leq C_{chl} \leq 1.0 \, (\mu\text{g/l})), \quad (4.18)$$

$$\log C_{sm} = 1.83 = 1.26 \log(X_{sm}) \quad (0.07 \leq C_{sm} \leq 0.56 \, (\text{mg/l})), \quad (4.19)$$

$$\log a_{doc}(440) = -3.00 - 1.93 \log(X_{doc}) \quad (0.01 \leq a_{doc}(440) \leq 0.065 \, (\text{m}^{-1})), \quad (4.20)$$

where $X_{chl} = [R(\lambda_2)/R(\lambda_5)][R(\lambda_1)/R(\lambda_3)]^{-1.2}$;

$\qquad X_{sm} = [R(\lambda_5)/R(\lambda_6)][R(\lambda_3)/R(\lambda_5)]^{-0.5}$;

$\qquad X_{doc} = [R(\lambda_1)/R(\lambda_3)][R(\lambda_2)]^{0.5}$,

λ_1–λ_6 are the wavelengths given for the SeaWiFS channels centered at: 412, 443, 490, 510, 555, 670 nm, respectively.

For waters rich in chlorophyll $(1 \leq C_{chl} \leq 40 \, \mu\text{g/l})$, suspended minerals $(0.56 \leq C_{sm} \leq 4.6 \, \text{mg/l})$, and $doc (a_{doc}(440) \sim 0.2 \, \text{m}^{-1})$, algorithms (4.18–4.20) have a slightly different form and also different regression coefficients:

$$\log C_{chl} = 0.36 - 4.38 \log(X_{chl}), \quad (4.21)$$

$$\log C_{sm} = 1.82 + 1.23 \log(X_{sm}), \quad (4.22)$$

$$\log a_{doc}(440) = -4.36 - 6.08 \log(X_{doc}). \quad (4.23)$$

Also the spectral reflection coefficient ratio products are different:

$\qquad X_{chl} = [R(\lambda_2)/R(\lambda_5)][R(\lambda_1)/R(\lambda_3)]^{-0.5}$;

$\qquad X_{sm} = [R(\lambda_5)/R(\lambda_6)][R(\lambda_3)/R(\lambda_5)]^{-1.2}$;

$\qquad X_{doc} = [R(\lambda_1)/R(\lambda_3)][R(\lambda_2)]^{0.25}$.

When choosing pairs of R relevant to various spectral bands of the SeaWiFS channels, Tassan (1994) searched for pairs with high sensitivity to the parameter to be retrieved and low sensitivity to fluctuations of the others. For example, for X_{chl} the wavelengths λ_2 and λ_5 are close to the maximum and minimum of chlorophyll absorption (i.e. the spectral influence of chlorophyll on $R(\lambda, -0)$ is the most pronounced), λ_1 and λ_3 are located on both sides of the chlorophyll absorption maximum. For X_{sm}, λ_5 and λ_6 correspond to a spectral region of weaker spectral influence (absorption) of phytoplankton *chl* and *doc* and to a high level of light scattering by suspended minerals, λ_3 and λ_5 are located between the spectral regions of strong *chl* and *doc* influence. For X_{doc}, λ_1 and λ_3 represent strong and weak absorption by *doc*, λ_2 compensates for *chl* influence.

The sensitivity of eqs (4.18–4.20) to fluctuations in concentrations of *doc* and *sm* relative to the respective values specified by the regression eqs (4.16) and (4.17) indicated that the retrieval errors did not exceed 20% to 30%. However, regardless of such optimistic assertions, it appears clear that the application of the above retrieval algorithms (4.18) to (4.20) to arbitrarily chosen Case II waters will unavoidably require a fine tuning of regression coefficients to local conditions as well as a thorough verification of retrieval results against relevant *in situ* truth data.

It should be pointed out, nevertheless, that the attempts to employ the band-ratio approach for Case II waters are not ceasing, and recently Ruddick *et al.* (2001) reported that it can be successfully improved by use of an adaptive two-band algorithm with optimal error properties.

Pozdnyakov and Lyaskovsky (1998) have carried out a numerical study aimed at assessing the robustness of the Tassan algorithms to a typical Case II inland water body, Lake Ladoga. Using the hydro-optical model developed for this lake (see Chapter 2), and the CPA concentrations typical of this water body, algorithms (4.18–4.20) were applied. The results in Fig. 4.2 indicate that the Tassan algorithms are strictly area-specific – either they are untranslatable to other basins or, rather, they rely on the assumption of the existence of strong inter-correlation between the principal CPAs in Case II waters that is often not valid.

In the case of eutrophic and hypertrophic waters, the development of chlorophyll retrieval algorithms is sometimes based on the use of only the red spectral region (Lin *et al.*, 1984). In the spectral distribution of $L_u(\lambda, +0)$ a minimum at about 670 nm, caused by the absorption of chlorophyll, is followed by a maximum between 690 and 720 nm (its actual position moves to longer wavelengths as chlorophyll concentration increases). This maximum is due to both fluorescence of chlorophyll and the enhancement of phytoplankton reflectance in the red and near IR portions of the spectrum (Gower and Borstad, 1993). The following regression algorithm based on the above combined effect has been suggested (Yacobi *et al.*, 1995):

$$C_{chl} = -A + B[L_u(\lambda_{max}, +0)/L_u(670\text{ nm}, +0)], \qquad (4.24)$$

where A and B are regression coefficients and λ_{max} is the wavelength of the red maximum in the spectral distribution of $L_u(\lambda, +0)$.

When hyperspectral remotely sensed data are available, the determination of C_{chl} could go via the integral value of the fluoresced radiance, if the spectrally integrated specific fluorescence of the phytoplankton assemblage Φ^*, $m^2 mg^{-1}$ is known. With known spectral distribution of fluorescence for individual species of phytoplankton (Φ_i^*) (see for example, Ahn *et al.*, 1992), it would be possible to infer the concentration of each component (C_i) of the phytoplankton assemblage via a multivariate optimization technique (for references see Kondratyev *et al.*, 1990) searching for the best fit between the measured and modelled radiance due to fluorescence.

The approach based on the integration procedure is also used to obtain regression relationships between spectrally integrated water surface reflectance (called spectral coefficient) and Secchi disk depth, S_D (Thiemann and Kaufmann, 2002). The exponential regression with a correlation coefficient $R_2 = 0.85$ and a mean error of 0.87 m revealed, as it should, an inverse relation between the spectral coefficient and S_D.

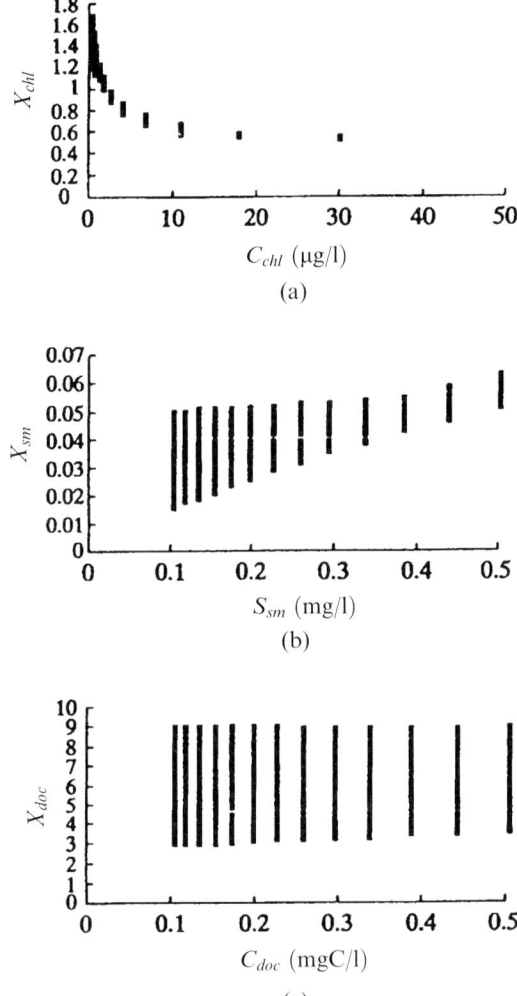

Figure 4.2. Results of regressing the concentrations of *chl* (a), *sm* (b), and *doc* (c) on parameters X_{chl}, X_{sm}, and X_{doc} respectively for Case II waters by applying the algorithms suggested by Tassan (1994).

Analysing the retrievals of C_{chl} from spectral radiances or reflectances in the red, it is worth mentioning that the $L_u(\lambda_{max}, +0)$ maximum in question is located within a region of fairly strong water absorption and absorption by atmospheric oxygen and water vapour. Therefore, the remotely recorded intensity of this signal is mainly due to planktonic cells residing in the near-surface layers. Therefore, the red spectrum algorithms are applicable only for highly productive waters with characteristic bloom events. In order to avoid interference by atmospheric absorption the small spectral bands have to be restricted to atmospheric windows.

Amongst the limitations inherent in these approaches within the red part of the visible spectrum is the diurnal variability in phytoplankton fluorescence (and this also concerns the *chl* fluorescence approach (Gower *et al.*, 1999) discussed above), whose neglect together with the effect of photoinhibition at high levels of incident radiation can seriously impact the results of chlorophyll content retrievals (for references see Kondratyev and Pozdnyakov, 1990b). Finally, since space sensors like CZCS and SeaWiFS record radiance signals not on a continuous basis, but only in several bands, and the spectral band 680–720 nm is not included, the red band-based approaches could not be applied for the currently operating space sensors, but they might be very useful for MERIS on ENVISAT.

The wish to retrieve *simultaneously* all major CPAs and not only *chl*, became an incentive of exploiting some other sophisticated mathematical methods beyond NNs. One of them is based on minimizing the χ^2 difference between the modelled and Rayleigh corrected remotely sensed radiances (Doerffer and Fischer, 1994). A two-flux radiative transfer model was calibrated with a set of radiance data which, in turn, were simulated with a *matrix-operator* radiative transfer model developed by Fischer and Doerffer (1987); see also Doerffer (1992). For modelling the light transfer through the water column, specific coefficients of absorption and attenuation a_i^*, c_i^* were used. A simplex optimization method (Nelder and Mead, 1965) was applied to minimize the value of χ^2 through varying the concentration vector C and the aerosol optical depth τ_a. When χ^2 becomes ≤ 0.3 (it requires about one hundred iterations), the corresponding values of C and τ_a are taken as the result of retrieval. This method has been tested with CZCS data over the North Sea. An important advantage of this approach is that it automatically allows for the atmospheric correction for radiances coming from Case II waters, also in the cases when the atmospheric correction algorithm suggested by Gordon and Wang (1994b) fails because of the absence of pixels for which $L_w(\lambda, +0) \approx 0$ at the long-wavelength channels of CZCS and SeaWiFS (see Chapter 5).

Another retrieval method, using a given set of absorption and backscattering cross-sections of CPAs, is based on *multivariate optimization* (described in detail in Kondratyev *et al.*, 1990; see also Roesler and Perry, 1995).

With such a hydro-optical model and an adequate parameterization of the volume reflectance $R(\lambda, -0)$, CPA concentrations are derived by the Levenberg–Marquardt multivariate optimization technique (Marquardt, 1963), the method of statistical regularization (Tikhonov and Arsonin, 1979), the maximum likelihood method (for references see Doubovick *et al.*, 1994), the linear matrix inversion technique (Hoge and Lyon, 1996), and others.

For the multivariate optimization technique, the volume reflectance just below the water surface, $R(\lambda, -0)$, has to be presented in vector form (Bukata *et al.*, 1985a,b):

$$R[\boldsymbol{a}^*(\lambda), \boldsymbol{b}_b^*(\lambda), \boldsymbol{C}] \sim \boldsymbol{b}_b^*(\lambda) \cdot \boldsymbol{C} / \boldsymbol{a}^*(\lambda) \cdot \boldsymbol{C}, \qquad (4.25)$$

where $\boldsymbol{a}^*(\lambda) = [a_w(\lambda), a_{chl}^*(\lambda), a_{sm}^*(\lambda), a_{doc}^*(\lambda), \ldots]$;

$\qquad \boldsymbol{b}_b^*(\lambda) = [b_{b_w}(\lambda), b_{b_{chl}}^*(\lambda), b_{b_{sm}}^*(\lambda), 0, \ldots]$;

$\qquad \boldsymbol{C} = (1, C_{chl}, C_{sm}, C_{doc}, \ldots)$.

With the measured spectrum S_j consisting of a set of volume reflectance values at discrete wavelengths λ_j, the weighted differences between the measured and modelled irradiance reflectance $R[a^*(\lambda), b_b^*(\lambda)C]$ can be written as

$$g_j(C) = \{S_j - R[a_j^*, b_{bj}^*, C]\}/S_j. \tag{4.26}$$

The multidimensional least-squares solution at the wavelengths corresponding to S_j is found by minimizing the residuals:

$$f(C) = \sum_j g_j^2(C). \tag{4.27}$$

The absolute minimum of $f(C)$ is found with the Levenberg–Marquardt finite difference algorithm (Marquardt, 1963; Levenberg, 1944). The result then is the concentration vector C for the location where the spectrum S_j has been recorded (a more detailed description of this method and technology of searching for the absolute minimum of $f(C)$ can be found in Kondratyev *et al.* (1990)).

An important advantage of such an approach is the simultaneous retrieval of all CPA concentrations included into the hydro-optical model. In addition, this approach is applicable to a wide range of the CPA concentrations, including waters with C_{doc} 5–10 g C/m^3 (Bukata *et al.*, 1985a). Also no correlation between the concentrations of major CPAs has to be assumed. This approach has been extensively and successfully used to retrieve CPA concentrations in Lakes Ontario, Ladoga, and Onega (Bukata *et al.*, 1985a,b; Kondratyev *et al.*, 1990).

The method of multivariate optimization can also be used to infer specific absorption and backscattering coefficients from remotely sensed reflectances or radiances (Bukata *et al.*, 1985a; Kondratyev *et al.*, 1990).

Recently Maritorena *et al.* (2002) suggested extending the Levenberg–Marquardt procedure with a procedure in which the simulated *annealing* technique is applied (Press *et al.*, 1992). Compared with other steepest descent minimization techniques that are intended to provide a rapid and precise solution, simulated annealing is an iterative heuristic method permitting you to search for the solution in the *uphill* direction. This increases the probability to find a *global* minimum. In addition, this approach reduces the importance of the first guess (see Kondratyev *et al.*, 1990) used to initiate the process of the residual minimization (eq. (4.27)) that is often a critical step in miminization techniques based on the steepest *descent* methods. Simulated annealing consists of three basic constituents: (1) a *cost* function that evaluates the performance of the procedure, (2) a candidate generator that randomly offers new values of the R vector, and (3) a decreasing temperature/cooling schedule that assures some randomness in the process and controls its overall performance. The candidate vectors are generated using the simplex method (Press *et al.*, 1992). This procedure allows the selection of an unfavourable solution (*uphill* step) that enables the optimization process to get out of *local* minima and eventually arrive at the *global* minimum. Importantly, the simulated annealing procedure has the ability to solve the inverse problem with more unknowns that the sole Levenberg–Marquardt method assures. This might be important, for instance, when retrieving bottom depth and *spectral* bottom albedo along with CPAs (see

Section 4.3.2). At the same time, this combined approach is far more time-consuming compared to the Levenberg–Marquardt procedure, which makes it more appropriate for determining hydro-optical parameters rather than for operational retrieval of water quality parameters from satellite images of non-Case I waters.

Returning to the Levenberg–Marquardt technique, a very similar approach has been developed by Roesler and Perry (1995): vectors of specific coefficients of absorption and backscattering could be presented as a sum of products,

$$\boldsymbol{a}(\lambda) = \sum_{i=1}^{I} a_i(\lambda) = \sum_{i=1}^{I} M_i \tilde{a}_i(\lambda), \tag{4.28}$$

$$\boldsymbol{b_b}(\lambda) = \sum_{j=1}^{J} b_{b_j}(\lambda) = \sum_{j=1}^{J} M_j \tilde{b}_{b_j}(\lambda), \tag{4.29}$$

where a_i and b_{b_j} are the coefficients of absorption and backscattering of the ith and jth CPAs, I and J are the total number of absorbing and scattering CPAs in the water column, M_i and M_j are the weights of absorption and backscattering coefficients of each component, and $\tilde{a}_i(\lambda)$ and $\tilde{b}_{b_j}(\lambda)$ are shape factors of ith and jth components, respectively.

The shape factors should be determined *a priori* very much like CPAs' specific absorption and backscattering coefficients: $a_i(\lambda)$ and $b_{b_j}(\lambda)$ from extensive observations are divided by their spectrally integrated values $(\tilde{a}_i(\lambda)$ and $\tilde{b}_{b_j}(\lambda))$ and then averaged. The retrieval procedure for vectors $\boldsymbol{a}, \boldsymbol{b_b}$ involving the multivariate optimization technique is in principle the same as the one described above for the retrieval of $\boldsymbol{C}$.

As can be seen in Section 4.2.1, neural networks are a promising tool for retrieving CPAs from volume reflectances or radiances. Schiller and Doerffer (1999) have reported their experience in creating a neural network specialized for solving the inverse problem in remote sensing of Case II waters. They showed that such a specialized network is capable of operationally retrieving the concentration vector $\boldsymbol{C}$ in hypothetical (modelled) waters even when very broad ranges of variations in an individual CPA are allowed.

Within the aforementioned comparative study (Pozdnyakov and Lyaskovsky, 1998) that was carried out from the perspective of the performance of some retrieval algorithms suggested in the literature for Case II waters, a neural network approach has been employed as well. Based on the Lake Ladoga hydro-optical model, a wide set of reflectance spectra has been generated to cover all realistic combinations of C_{chl}, C_{sm}, and C_{doc}. These data sets were then used to assess the efficiency of both the multivariate optimization and neural network techniques.

Following Schiller and Doerffer (1999), to approximate the inverse model a feed-forward (error back-propagation) network was chosen (the Stuttgart University Neural Network (SNNS, 1995)). The neural network was composed of four layers of 'neurons': an input layer, two hidden layers, and an output layer. The number of units in the input layer equals the number of input values (i.e. six reflectances at the SeaWiFS wavelengths in the visible spectrum). The output layer parameter number corresponds to the concentrations of major optically active components (*chl*, *sm*, *doc*) and the parameter number in the hidden layer is problem-dependent. An optimum performance was found for six input units, twenty in the first, five in the second hidden layer, and three in the output layer.

The training data, generated from absorption and backscattering cross-sections for Lake Ontario and the *R* formulation developed by Jerome *et al.* (1988b), was presented to the designed neural network in a random order. The error minimization was continued until the error function corresponded to an average error of 1.2%. At this level the configuration of the neural network was considered appropriate for conducting retrieval experiments with the generated data.

The retrieval comparison showed that the best results are obtained with the Levenberg–Marquardt method (Fig.4.3).

The neural network approach delivered less accurate results in comparison with the Levenberg–Marquardt method (Fig. 4.4). However, this might be the result of a non-optimal configuration designed in this study, although many options were tested and the used one proved to be the best. The NN algorithm robustness might be further enhanced if the range of CPA concentrations is narrowed. Such specialized neuron networks for predetermined narrow concentration ranges could be generated for distinct water bodies or areas.

4.3 INFERRING CPAs FROM WATER COLOUR FOR OPTICALLY DEEP AND SHALLOW WATERS WITH THE MULTIVARIATE OPTIMIZATION TECHNIQUE: NUMERICAL SIMULATIONS

4.3.1 Deep/semi-infinite waters

In accordance with the conceptual approach outlined in Section 4.1, a remotely sensed colour proxy, like spectral volume reflectance $R(\lambda, -0)$ is subjected to a two-step process.

Firstly, CPAs are retrieved assuming that the co-existing CPAs only absorb and scatter. This could be done using (i) one of the parameterizations relating $R(\lambda, -0)$ to the bulk absorption and backscattering coefficients, $a(\lambda, C), b_b(\lambda, C)$, where C, as above, is the concentration vector of co-existing CPAs; (ii) a hydro-optical model relating, through the appropriate cross-sections, the bulk absorption and backscattering coefficients $a(\lambda, C), b_b(\lambda, C)$ to the concentrations of the co-existing CPAs; and (iii) a retrieval procedure, e.g. the Levenberg–Marquardt multivariate optimization procedure, or neural networks or any other one with comparable capabilities (see Section 4.2) to obtain the first guess CPA concentration vector $C_{fg}(C_{chl}, C_{sm}, C_{doc})$ in the case of a four-component hydro-optical model.

Secondly, based on this first guess concentration vector C_{fg}, (i) assess the contributions to $R(\lambda, -0)$ due to other photon interactions with the aquatic medium, viz., water Raman scattering, fluorescence by chlorophyll, fluorescence by dissolved organic matter $[R_r(\lambda, -0), R_{chl}^f(\lambda, -0), R_{doc}^f(\lambda, -0)$, respectively; (ii) subtract the

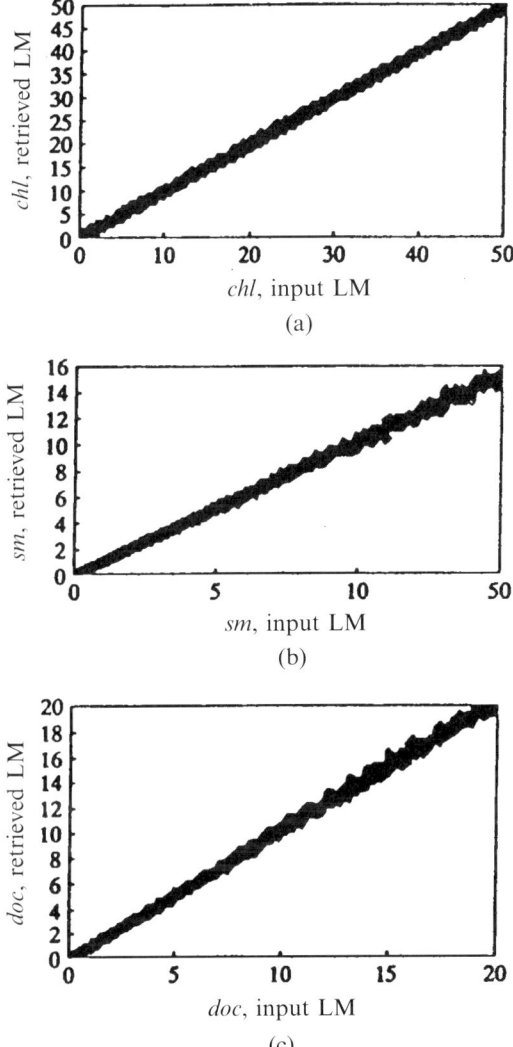

Figure 4.3. Performance of the Levenberg–Marquardt procedure for the simultaneous retrieval of *chl* (a), *sm* (b), and *doc* (c) concentrations for Case II waters.

spectral values of $R_r(\lambda, -0)$, $R^f_{chl}(\lambda, -0)$, $R^f_{doc}(\lambda, -0)$, from $R(\lambda, -0)$; and (iii) apply the retrieval procedure to find a new concentration vector C. If the procedure is converging, this C should be more accurate.

In view of the impact of trans-spectral processes (see Chapter 3), neglecting $R_r(\lambda, -0)$, $R^f_{chl}(\lambda, -0)$, $R^f_{doc}(\lambda, -0)$ also will lead to errors in the retrievals of the concentration vector C.

Pozdnyakov *et al.* (2002a) conducted numerical simulations to quantitatively assess these errors, and, if necessary to apply an iterative scheme for corrections.

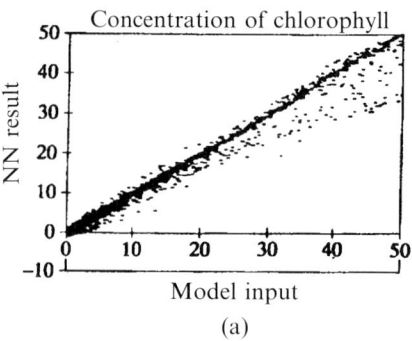

(a)

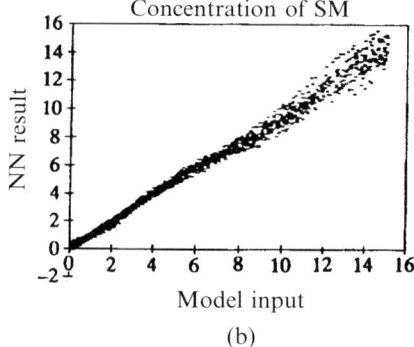

(b)

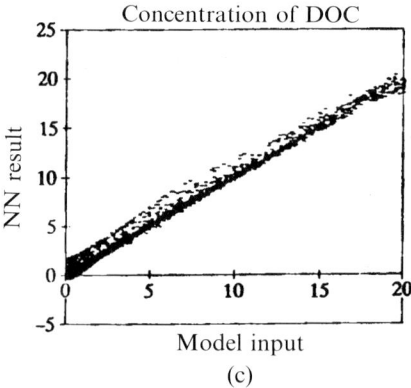

(c)

Figure 4.4. Performance of the neural network approach for the simultaneous retrieval of *chl* (a), *sm* (b), and *doc* (c) concentrations for Case II waters.

Similarly to the forward problem (Chapter 3), only the nadir view was considered here. Calculations were again performed only for a calm water surface and the following depth-independent concentrations: C_{chl} (µg/l) $= 0.0, 0.5, 1.0, 2.0, 3.0, 4.0,$ $5.0, 15.0$; C_{sm} (mg/l) $= 0.0, 0.5, 1.0, 2.0, 3.0, 4.0, 5.0$; C_{doc} (mg C/l) $= 0.0, 0.5, 1.0, 2.0,$ $5.0, 10.0$; whereby the trophic status ranges from oligotrophic to meso-eutrophic

(Petrova, 1990). All calculations were conducted for solar zenith angles equal to $0°$, $5°$, $10°$, $30°$, $40°$, and $50°$. The Lake Ladoga hydro-optical model with four CPAs was applied (see Tables 2.1 and 2.2). Eqs (1.18), (1.19), (1.36) and (1.40) were used to model $R(\lambda, -0)$, and $R_r(\lambda, -0)$, $R_{chl}^f(\lambda, -0)$, $R_{doc}^f(\lambda, -0)$, respectively. The values of other input parameters (e.g. $\eta_{chl}, \eta_{doc}(\lambda)$) required for calculating $R_r(\lambda, -0)$, $R_{chl}^f(\lambda, -0)$, $R_{doc}^f(\lambda, -0)$ are defined in Sections 3.1.1 to 3.1.3. As expected, the concentration vector retrieval accuracy is sufficiently high only if the fluorescence yield both for *chl* and *doc* is low. In Table 4.1 it is illustrated for two values of η_{chl}: 0.7 and 3%. The fluorescence-driven contributions to volume reflectance $R(\lambda, -0)$ can be considered as perturbations to the input parameters within the Levenberg–Marquardt retrieval procedure, which are the measured/determined spectrum of $R(\lambda, -0)$, and/or the applied hydro-optical model. Indeed, it was shown (Pozdnyakov and Lyaskovsky, 1999) that the retrieval accuracy deteriorates dramatically only when the perturbations to input parameters are in excess of $\sim 5\%$.

To characterize the retrieval errors arising from the neglect of transspectral processes, the following ratios were calculated:

$$\Delta C_{doc} = (C_{doc}^{input} - C_{doc}^{retrieved})/C_{doc}^{input},$$

$$\Delta C_{sm} = (C_{sm}^{input} - C_{sm}^{retrieved})/C_{sm}^{input},$$

$$\Delta C_{chl} = (C_{chl}^{input} - C_{chl}^{retrieved})/C_{chl}^{input},$$

where C_i^{input} is the concentration of an optical constituent used in the calculations. The simulations indicate that, given that $\eta_{doc}(\lambda)$ and $\eta_{chl} = 0.7\%$, the retrieval error ΔC_{doc} in the case of water devoid of phytoplankton and suspended minerals is always positive, increasing up to $\sim 57\%$ at $C_{doc} = 10\,\text{mg}\,\text{C/l}$ (Table 4.2), which implies that the retrieved C_{doc} is *underestimated*.

However, addition of chlorophyll ($\leq 5\,\mu\text{g/l}$) results in a rapid decrease of ΔC_{doc} to 20% and even less (Table 4.3). A further growth of chlorophyll content leads to *overestimation* of C_{doc}, and ΔC_{doc} exceeds 50% at $C_{chl} = 15\,\mu\text{g/l}$. Additionally, these variations in ΔC_{doc} change strongly if water contains not only chlorophyll but also suspended minerals. For $C_{chl} \leq 5\,\mu\text{g/l}$, the addition of suspended minerals enhances the ΔC_{doc} reduction, whereas for $C_{chl} > 5\,\mu\text{g/l}$, the effect is opposite, i.e. C_{doc} decreases with increasing concentrations of both chlorophyll and suspended minerals.

The analyses of retrieval errors for *chl* indicate that (i) the inaccuracy in inferring C_{chl} increases with C_{doc}, and (ii) it is strongly dependent on the concentration of suspended minerals: the more suspended minerals, the lower ΔC_{chl}, which is illustrated in Table 4.4.

In turn, as it is exemplified in Table 4.5, the C_{sm} retrieval accuracy depends on the fluorescence of chlorophyll and dissolved organics. For $C_{chl} \leq 5\,\mu\text{g/l}$ and $C_{doc} \leq 2\,\text{mg}\,\text{C/l}$, ΔC_{sm} can be both positive and negative, not exceeding, however, 30%. If both C_{doc} and C_{chl} increase up to $10\,\text{mg}\,\text{C/l}$ and $15\,\mu\text{g/l}$, respectively, ΔC_{sm} progressively decreases: the highest *overestimation* of C_{sm} occurs at lowest C_{doc}.

Table 4.1. Impact of Raman scattering by water molecules, chlorophyll fluorescence and dissolved organics fluorescence on the retrieval of concentrations (in round figures) for some combinations of CPA concentrations, $\theta_0 = 30°$, and chlorophyll fluorescence quantum yields 0.7% and 3%.

Vector	chl, μg/l		sm, mg/l		doc, mgC/l	
	$\eta_{chl} = 0.7\%$	$\eta_{chl} = 3\%$	$\eta_{chl} = 0.7\%$	$\eta_{chl} = 3\%$	$\eta_{chl} = 0.7\%$	$\eta_{chl} = 3\%$
Input	5		0		0	
Retrieved	6.0	4.3	0.0	0.5	0.1	1.8
Input	5		0		0.1	
Retrieved	6.5	2.5	0.0	0.6	0.1	2.7
Input	5		0		1	
Retrieved	5.3	0.4	0.1	0.1	1.3	5.6
Input	5		0		2	
Retrieved	0.7	2.8	0.2	4.7	2.4	0.0
Input	5		0		5	
Retrieved	0.0	15.1	0.3	0.3	4.5	11.0
Input	0		5		0	
Retrieved	0.4	0.2	5.3	5.8	0.0	0.2
Input	5		0.1		0	
Retrieved	5.2	5.5	0.2	0.5	0.1	1.0
Input	5		1		0	
Retrieved	4.7	5.1	1.1	1.5	0.1	0.3
Input	5		2		0	
Retrieved	5.1	5.0	2.1	2.6	0.0	0.3
Input	5		5		0	
Retrieved	4.4	5.0	5.1	5.9	0.1	0.2
Input	5		2		0.1	
Retrieved	5.0	5.0	2.1	2.6	0.1	0.4
Input	5		2		1	
Retrieved	4.6	4.1	2.1	2.6	1.0	1.6
Input	5		2		2	
Retrieved	4.5	3.3	2.1	2.6	2.0	2.7
Input	5		2		5	
Retrieved	0.5	1.0	2.1	2.5	5.0	5.5
Input	0		0		0.1	
Retrieved	0.1	0.1	0.0	0.0	0.1	0.1

Table 4.2. Relative retrieval error, ΔC_{doc}, if the fluorescence of dissolved organic matter is neglected. $C_{chl} = C_{sm} = 0, \theta_0 = 30°$.

	C_{doc}, mg C/l				
	0.5	1.0	2.0	5.0	10.0
ΔC_{doc}	0.27	0.34	0.41	0.51	0.57

Table 4.3. Relative retrieval error, ΔC_{doc}, if the fluorescence of dissolved organic matter and phytoplankton is neglected, $C_{doc}^{input} = 2$ mg C/l, $\theta_0 = 30°$.

	η_{chl}, %	C_{sm}, mg/l	C_{chl}, µg/l						
			0.5	1.0	2.0	3.0	4.0	5.0	15.0
ΔC_{doc}	0.7	0.5	0.23	0.13	−0.04	−0.07	−0.10	−0.13	−0.53
		1.0	0.12	0.09	−0.01	−0.02	−0.04	−0.07	−0.37
		2.0	0.11	0.04	0.03	0.01	−0.01	−0.03	−0.24
		3.0	0.14	0.04	0.03	0.02	0.01	−0.05	−0.18
		4.0	0.16	0.05	0.03	0.02	0.01	−0.00	−0.15
		5.0	0.18	0.05	0.04	0.03	0.01	0.00	−0.12
		0.5	0.09	0.05	−0.22	−0.45	−0.68	−0.89	−4.48
		1.0	0.09	0.00	−0.16	−0.28	−0.42	−0.56	−3.07
	3.0	2.0	0.10	0.00	−0.08	−0.16	−0.25	−0.34	−1.98
		3.0	0.11	0.00	−0.06	−0.12	−0.18	−0.25	−1.42
		4.0	0.15	0.01	−0.04	−0.09	−0.15	−0.20	−0.14
		5.0	0.20	0.01	−0.03	−0.08	−0.12	−0.18	−0.98

Table 4.4. Relative retrieval error, ΔC_{chl}, if the fluorescence of dissolved organic matter and phytoplankton is neglected, $C_{chl}^{input} = 2$ µg/l, $\theta_0 = 30°$.

	η_{chl}, %	C_{sm}, mg/l	C_{doc}, mg C/l				
			0.5	1.0	2.0	5.0	10.0
ΔC_{chl}	0.7	0.5	0.17	0.37	0.65	0.95	0.97
		1.0	0.12	0.25	0.48	0.95	0.97
		2.0	0.08	0.17	0.34	0.86	0.80
		3.0	0.07	0.14	0.28	0.76	0.87
		4.0	0.06	0.12	0.25	0.69	0.95
		5.0	0.05	0.11	0.23	0.64	0.78
		0.5	0.35	0.66	0.99	0.97	0.90
		1.0	0.25	0.43	0.73	0.97	0.84
	3.0	2.0	0.18	0.29	0.49	0.91	0.89
		3.0	0.15	0.23	0.40	0.92	0.99
		4.0	0.13	0.20	0.35	0.82	0.93
		5.0	0.12	0.18	0.31	0.75	0.89

Table 4.5. Relative retrieval error, ΔC_{sm}, if the fluorescence of dissolved organic matter and phytoplankton is neglected. $C_{sm}^{input} = 3$ mg/l, $\theta_0 = 30°$.

	η_{chl}, %	C_{doc}, mg C/l	C_{chl}, µg/l						
			0.5	1.0	2.0	3.0	4.0	5.0	15.0
		0.5	−0.01	−0.02	−0.03	−0.04	−0.05	−0.07	−0.23
		1.0	−0.02	−0.02	−0.03	−0.04	−0.05	−0.07	−0.22
ΔC_{sm}	0.7	2.0	−0.11	−0.01	−0.02	−0.03	−0.04	−0.06	−0.21
		5.0	−0.21	−0.15	−0.11	0.03	0.02	0.00	−0.12
		10.0	−0.23	−0.25	0.00	−0.01	0.03	0.04	0.02
	3.0	0.5	−0.03	−0.06	−0.11	−0.17	−0.23	−0.30	−1.57
		1.0	−0.03	−0.06	−0.11	−0.17	−0.23	−0.30	−1.54
		2.0	−0.12	−0.05	−0.10	−0.16	−0.22	−0.29	−1.46
		5.0	−0.08	−0.13	−0.23	−0.16	−0.15	−0.21	−1.18
		10.0	−0.06	−0.05	−0.12	−0.16	−0.19	−0.41	−0.77

As can be seen from Table 4.5, within the explored ranges of concentrations and with $\eta_{chl} = 0.7\%$, C_{sm} is likely to be slightly *overestimated* (by ~1.5%) at low values of both C_{doc} and C_{chl} and more substantially *overestimated* (by ~25%) at high values of C_{chl} and low values of C_{doc}.

In summary. The neglect of trans-spectral processes has serious concequences:

- C_{doc} is *underestimated* up to ~25% at low values of both C_{chl} and C_{sm}, and even up to ~50% at $C_{chl} = C_{sm} = 0$ (Table 4.3).
- C_{chl} is *underestimated* at any combination of C_{doc} and C_{sm}: up to 100% at high values of C_{doc} and low values of C_{sm} (Table 4.4).
- C_{sm} is mostly slightly *overestimated* at low C_{doc} and C_{chl} but substantially *overestimated* (~25%) for high C_{doc} and C_{chl}.

The retrieval errors grow with η_{chl} particularly for dissolved organics and suspended minerals: at high concentrations of chlorophyll and low concentrations of suspended minerals, C_{doc} may be *overestimated* even by a factor of ~5 (Table 4.3), whereas at high C_{chl} and low C_{doc}, the content of suspended minerals may be *overestimated* by up to a factor of ~1.5 (Table 4.5). Although the relative error of chlorophyll retrieval ΔC_{chl} seems to be less sensitive to the value of η_{chl}, it, the *underestimation*, nevertheless, increases with η_{chl}, being especially strong at high C_{doc} (Table 4.4).

These results suggest that adequate CPA concentration retrievals are most likely only possible if the fluorescence yields of both chlorophyll and dissolved organics are known with reasonable confidence. Thus further extensive laboratory and field experiments, conducted for very different compositions of coastal and inland waters, are necessary.

Thus we emphasize again that fluorescence in natural waters strongly affects the retrievals, if the latter are calculated with algorithms/procedures derived without due reference to trans-spectral effects. In addition, the degradation of the retrieval accuracy due to the neglect of fluorescence by chlorophyll and dissolved organics is a strong function of the CPA concentration vector. Therefore, advanced hydro-optical models need tabulated spectral values of the fluorescence yields η_{chl} and η_{doc} as input. As both absorption and scattering cross-sections of CPAs as well as fluorescence yields of *chl* and *doc* are variable and water body- and season-specific, more complexity is thus added.

Consequently, a further improvement of the accuracy of water quality retrievals needs hydro-optical models as a function of major vegetation seasons and water bodies. However, it is not unreasonable to expect (see, for example, Pozdnyakov *et al.*, 1999; Bukata *et al.*, 2001) that the variety of such area- and season-specific models can be reduced to some typical ones (in terms of latitudinal zone, vegetation season, landscape type, anthropogenic loading), which can be applied either alone or in combinations to different water bodies.

Notwithstanding a certain complication of the retrieval procedure, the impact of trans-spectral processes can be handled through an iterative procedure outlined at the beginning of this section. Fig. 4.5 depicts such a two-step scheme.

Table 4.6 provides an assessment of the efficiency of the scheme in Fig. 4.5 designed to improve the accuracy of CPA retrievals. The results listed in Table 4.6 are based on $\eta_{chl} = 0.7\%$ and $\eta_{doc}(\lambda)$ as specified in Chapter 3 (Section 3.1.3). As can be seen, in the first step the highest error arises for the retrieval of C_{chl}, whereas C_{sm} and C_{doc} are inferred rather accurately. Owing to rapid convergence of the procedure, by the second step there already results a significant improvement of the C_{chl} retrieval, also accompanied by higher accuracy of C_{sm} and C_{doc} determinations.

4.3.2   Optically thin/shallow waters

Estep (1994) investigated by means of modelling the impact of bottom albedo on the retrieval of water quality parameters from remote sensing (see, for example, eq. (4.1)). He showed that a depth $H = 5\,m$, in Case I waters with chlorophyll content ranging from 0 to $3\,\mu g/l$, leads to large retrieval errors up to an order of magnitude for different bottom covers like sand, silt, or green algae.

For input *chl* concentrations confined between $10\,\mu g/l$ and $20\,\mu g/l$, the C_{chl} obtained from eq. (4.1) can even be (for Case I waters!) as high as $35-85\,\mu g/l$, such high concentrations being characteristic of hypertrophic waters. However, for $H = 10\,m$, the error due to the impact of bottom albedo considerably decreases and becomes negligible at $H = 20\,m$.

The influence of bottom albedo in Case II waters that are often relatively opaque for solar radiation, can be traced only for depths essentially smaller than those cited above.

When remotely sensing optically shallow waters, not only CPA concentrations but also bottom *depth* and bottom *albedo* have to be included in the hydro-optical model.

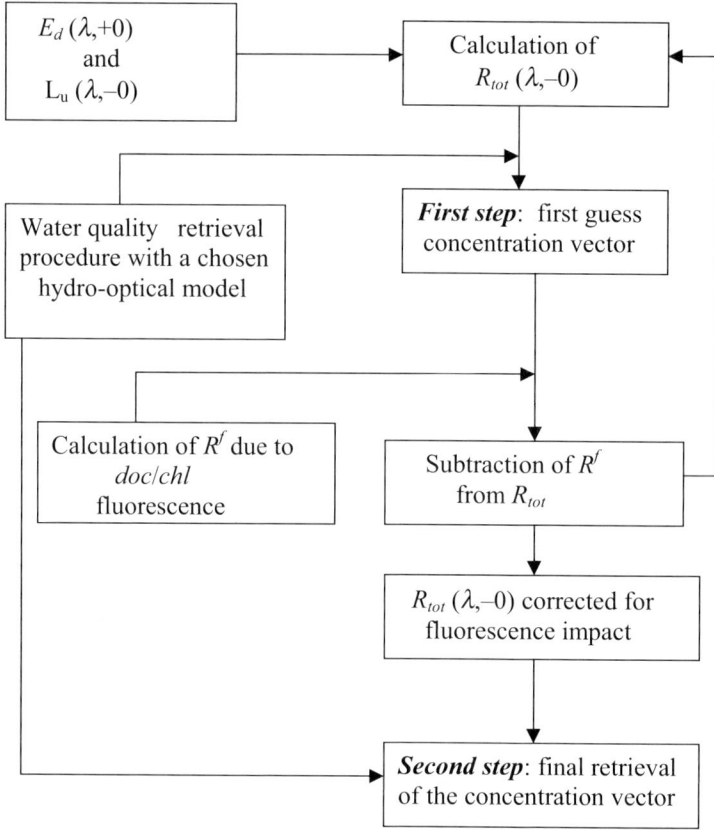

Figure 4.5. A two-step scheme developed for improving the accuracy of retrieval of water quality parameters by taking into account trans-spectral processes.

Table 4.6. Results of application of the two-step procedure for improving the retrievals of CPAs (C_{input} is the CPA concentration input; $C_{ret,I}$, $C_{ret,II}$ are the retrieval results obtained with the use of the Levenberg–Marquardt algorithm for the first and second step).

Type of the concentration vector $C(C_{chl}, C_{sm}, C_{doc})$	C_{chl}, µg/l	C_{sm}, mg/l	C_{doc}, mg C/l
C_{input}	2.0	0.5	0.5
$C_{ret,I}$	1.5	0.6	0.6
$C_{ret,II}$	2.1	0.5	0.5
C_{input}	2.0	0.5	2.0
$C_{ret,I}$	0.2	0.6	2.1
$C_{ret,II}$	2.1	0.5	2.1
C_{input}	2.0	0.5	10.0
$C_{ret,I}$	0.7	0.7	9.2
$C_{ret,II}$	2.2	0.5	10.5

The traditional regression technique has also been and is still widely used for shallow waters. As a recent example, Lafon *et al.* (2002) have suggested for moderately turbid waters of the inlet of Arcachon lagoon a regression relating the bottom depth and the so called *in situ* reflectance $R_{is} = \pi R_{rs}$ measured in the first SPORT channel (500–560 nm). The established regression is valid for $C_{sm} \leq 9.0$ mg/l and bottom depths less than 6 m.

Based on their earlier studies (Lee *et al.*, 1998, 1999) and applying their computer optimization code as well as the previously developed hydro-optical model (Lee *et al.*, 2001), Lee and Carder (2002) succeeded in retrieving from the spectral values of R_{rs}, in addition to bulk water absorption at 440 nm, phytoplankton pigment and *gelbstoff* (yellow substance) absorption at the same wavelength, particle backscattering coefficient, at 640 nm, as well as the bottom depth and bottom albedo (at 550 nm) for a variety of hydro-optical conditions typical of oceanic costal waters. It was also shown that since the detection of bottom depth strongly depends on the in-water properties, the spectral channels of remote sensors should match the water's spectral transparency window, which generally ranges in clear/slightly turbid coastal waters from 470 to 570 nm. The limited number of bands available from SeaWiFS and MODIS does not necessarily ensure the detection of the strongest bottom signals given the variety of hydro-optical conditions, bottom topography and bottom reflectivity properties indigenous to coastal environments.

Analysing the inverse problem for optically shallow waters, Pozdnyakov *et al.* (2001) have modelled the following hydro-optical conditions: calm water surface, depth-independent concentrations $C_{chl}(\mu g/l) = 0.0, 1.0, 5.0, 10.0, 20.0$; $C_{sm}(mg/l) = 0.0, 0.5, 5.0$; $C_{doc}(mg\,C/l) = 0.0, 0.1, 1.0, 5.0, 10.0, 15.0$; solar zenith angles from $0°$ to $50°$; $H(m) = 1.0, 5.0, 10.0, 20.0, 50.0$.

The Lake Ladoga hydro-optical model was employed in its extended version (see Tables 2.1 and 2.2) with eqs (1.18), (1.19), (1.36) and (1.40), thus modelling $R(\lambda, -0)$, and $R_r(\lambda, -0)$, $R_{chl}^f(\lambda, -0)$, $R_{doc}^f(\lambda, -0)$, respectively. $\eta_{chl}, \eta_{doc}(\lambda)$, and other parameters required for $R_r(\lambda, -0)$, $R_{chl}^f(\lambda, -0)$, $R_{doc}^f(\lambda, -0)$ are as defined in Sections 3.1.1–3.1.3. The Kirk parameterizations for $K_{sun}(\lambda, \theta_0')$ and $K_{sky}(\lambda)$ (eqs (1.26) and (1.27)) were used in conjunction with expressions (1.28) and (1.29).

Following Maritorena *et al.* (1994) and Estep (1994), $R_{tot}(\lambda, -0, H, C)$ was calculated from eq. (1.46) with the spectral albedo for silt and the alga *Boodlea* (see Fig. 3.21).

The Levenberg–Marquardt multivariate optimization retrieval procedure with a slightly modified form of eq. (4.26) was used:

$$g_j(\mathbf{C}, H) = (S_j - R_{tot,j}(\mathbf{C}, H))/S_j. \qquad (4.30)$$

The task is then reduced to jointly determine concentration vector $\mathbf{C}$ and bottom depth H that provide a global minimum of the squared sum of residuals:

$$f(\mathbf{C}, H) = \sum_j g_j^2(\mathbf{C}, H). \qquad (4.31)$$

Thus, this approach allows us to *simultaneously* restore CPA concentrations and bottom depth H, i.e. *four* variables.

The application of the Levenberg–Marquardt technique to $R_{tot}(\lambda, -0.H, C)$ excludes bottom influence, not only in mesotrophic/fairly turbid but also in oligotrophic/transparent waters. Table 4.7 contains, for a bottom covered by *Boodlea*, the retrievals of *chl*, *sm*, and *doc* from $R_{tot}(\lambda, -0, H, C)$ at $\theta_0 = 30°$, if the transspectral processes are neglected. The concentration vector $C(C_{chl}, C_{sm}, C_{doc})$ can be restored with high accuracy within the bottom depth range (1–20) m, at a given spectral bottom albedo. The retrieval of bottom depth H strongly depends on the CPA concentration vector and on H itself. For $H \leq 10$ m and $C_{chl} < 5$ µg/l, the accuracy of the retrieval of H is high, as it is for C. However, as C_{chl} moves into the range of 10 to 20 µg/l, and $H \geq 10$ m, it can no longer be retrieved. When H approaches 20 m, it already differs substantially from the 'true' value for $C_{chl} = 0.1$ µg/l, $C_{sm} \cong 0.5$ mg/l, and $C_{doc} \cong 1.0$ mgC/l.

Modelling eutrophic waters and waters with enhanced turbidity (see Table 4.8 for a solar zenith angle $\theta_0 = 30°$ and the bottom covered with the alga *Boodlea*) gives high accuracy in the retrieval of the concentration vector $C(C_{chl}, C_{sm}, C_{doc})$, fairly independent of θ_0 and bottom cover (silt or green algae). However, H cannot be retrieved.

Inclusion of trans-spectral processes reduces retrieval accuracy both for the concentration vector C and bottom depth H. This is illustrated in Table 4.9 for a bottom covered by *Boodlea* at $\theta_0 = 30°$.

The two-step procedure helps in clear and really shallow waters ($H = 1$ m), but the retrieval results worsen for less transparent/more turbid and deeper waters. This underlines the difficulties of remote sensing in Case II/coastal waters at depths where bottom albedo still matters.

4.4 ACCOUNTING FOR THE CONTRIBUTION OF SURFACE WAVES TO THE WATER-LEAVING RADIANCE

So far, we have discussed the problem of extracting the concentration of CPAs in natural water bodies under the assumption of a calm air–water interface. However, in reality this is an exceptional rather than a regular case, and surface roughness effects contribute to total radiance $L_{u,tot}(\lambda, +0, \theta_v, \phi_v)$ leaving the water–air interface in the viewing direction θ_v, ϕ_v of a remote sensor (Krotkov and Vasilkov, 2000). Taking this into account presents a certain challenge.

Following Bukata *et al.* (1995), consider the transfer of $E_u(\lambda, -0)$ through the water–air interface and its contribution to $L_{u,tot}(\lambda, +0, \theta_v, \phi_v)$:

$$L_{u,tot}(\lambda, +0, \theta_v, \phi_v) = f_1 E_{sky}(\lambda) + f_2 E_{sun}(\lambda, \theta_0) + E_u(\lambda, -0) T_{surf}/Q, \qquad (4.32)$$

where $E_{sky}(\lambda), E_{sun}(\lambda, \theta_0)$ are the incident diffuse and direct solar irradiances, respectively (for sun zenith angle θ_0); Q is defined following eq. (1.20): $Q = E_u(\lambda, -0)/L_u(\lambda, -0, 0°, \phi_v)$; T_{surf} is the transmission of nadir radiance through the water–air interface; f_1 is the portion of $E_{sky}(\lambda)$ reflected at the water surface into the field-of-view of a remote sensor; and f_2 is the portion of $E_{sun}(\lambda, \theta_0)$ reflected at the water surface into the field-of-view of a remote sensor. For simplicity of notation we

Table 4.1. Results of retrieval of C_{chl} (µg/l), C_{sm} (mg/l), C_{doc} (mg/l), and H (m). Parameters without an asterisk are the input parameters; parameters with an asterisk are the retrieved ones. X stands for the combinations of CPAs simulated in each concrete case. Trans-spectral processes are excluded.

C_{chl}	C_{sm} 0.1	C_{sm} 0.5	C_{sm} 1.0	C_{doc} 0.1	C_{doc} 1.0	C_{chl}^*	C_{sm}^*	C_{doc}^*	H^*	C_{chl}^*	C_{sm}^*	C_{doc}^*	H^*	C_{chl}^*	C_{sm}^*	C_{doc}^*	H^*	C_{chl}^*	C_{sm}^*	C_{doc}^*	H^*
							$H=1$				$H=5$				$H=10$				$H=20$		
0.1	X			X		0.1	0.1	0.1	1.0	0.1	0.1	0.1	5.0	0.1	0.1	0.1	10.0	0.1	0.1	0.1	19.9
	X				X	0.1	0.1	1.0	1.0	0.1	0.1	1.0	5.0	0.2	0.1	1.0	10.0	0.2	0.1	1.0	20.9
		X		X		0.1	0.5	0.1	1.0	0.1	0.5	0.1	5.0	0.1	0.5	0.1	10.0	0.1	0.5	0.1	20.0
		X			X	0.1	0.5	1.0	1.0	0.1	0.5	1.0	5.0	0.1	0.5	1.0	10.0	0.1	0.5	1.0	64.2
			X	X		0.1	1.0	0.1	1.0	0.1	1.0	0.1	5.0	0.1	1.0	0.1	10.0	0.1	1.0	0.1	21.0
			X		X	0.1	1.0	1.0	1.0	0.1	1.0	1.0	5.0	0.1	1.0	1.0	10.0	0.1	1.0	1.0	48.0
1.0	X			X		1.0	0.1	0.1	1.0	1.0	0.1	0.1	5.0	1.0	0.1	0.1	10.0	1.0	0.1	0.1	19.9
	X				X	1.0	0.1	1.0	1.0	1.0	0.1	1.0	5.0	1.1	0.1	1.0	10.0	1.1	0.1	1.0	62.6
		X		X		1.0	0.5	0.1	1.0	1.0	0.5	0.1	5.0	1.0	0.5	0.1	10.0	1.0	0.5	0.1	20.3
		X			X	1.0	0.5	1.0	1.0	1.0	0.5	1.0	5.0	1.1	0.5	1.0	10.0	1.1	0.5	1.0	54.7
			X	X		1.0	1.0	0.1	1.0	1.0	1.0	0.1	5.0	1.0	1.0	0.1	10.0	1.0	1.0	0.1	66.7
			X		X	1.0	1.0	1.0	1.0	1.0	1.0	1.0	5.0	1.0	1.0	1.0	10.1	1.0	1.0	1.0	45.1
5.0	X			X		5.0	0.1	0.1	1.0	5.0	0.1	0.1	5.0	5.1	0.1	0.1	10.1	5.2	0.1	0.1	78.4
	X				X	5.0	0.1	1.0	1.0	5.1	0.1	1.0	5.0	5.2	0.1	1.0	10.2	5.1	0.1	1.0	52.8
		X		X		5.0	0.5	0.1	1.0	5.1	0.5	0.1	5.0	5.1	0.5	0.1	10.1	5.1	0.5	0.1	61.2
		X			X	5.0	0.5	1.0	1.0	5.1	0.5	1.0	5.0	5.1	0.5	1.0	10.4	5.1	0.5	1.0	47.2
			X	X		5.0	1.0	0.1	1.0	5.1	1.0	0.1	5.0	5.1	1.0	0.1	10.2	5.1	1.0	0.1	54.7
			X		X	5.0	1.0	1.0	1.0	5.1	1.0	1.0	5.0	5.1	1.0	1.0	11.6	5.1	1.0	1.0	39.3
10.0	X			X		10.0	0.1	0.1	1.0	10.1	0.1	0.1	5.0	10.2	0.1	0.1	10.3	10.1	0.1	0.1	57.0
	X				X	10.1	0.1	1.0	1.0	10.2	0.1	1.0	5.0	10.3	0.1	1.0	39.2	10.1	0.1	1.0	42.5
		X		X		10.0	0.5	0.1	1.0	10.1	0.5	0.1	5.0	10.2	0.5	0.1	10.6	10.1	0.5	0.1	49.4
		X			X	10.1	0.5	1.0	1.0	10.2	0.5	1.0	5.0	10.1	0.5	1.0	37.3	10.1	0.5	1.0	37.0
			X	X		10.1	1.0	0.1	1.0	10.1	1.0	0.1	5.0	10.1	1.0	0.1	14.9	10.1	1.0	0.1	40.3
			X		X	10.1	1.0	1.0	1.0	10.2	1.0	1.0	5.0	10.1	1.0	1.0	33.3	10.1	1.0	1.0	33.3
20.0	X			X		20.1	0.1	0.1	1.0	20.4	0.1	0.1	5.1	20.1	0.1	0.1	35.2	20.1	0.1	0.1	36.0
	X				X	20.1	0.1	1.0	1.0	23.4	0.0	0.7	17.4	20.1	0.1	1.0	29.2	20.1	0.1	1.0	29.2
		X		X		20.1	0.5	0.1	1.0	20.4	0.5	0.1	5.1	20.1	0.5	0.1	33.3	20.1	0.5	0.1	33.2
		X			X	20.1	0.5	1.0	1.0	21.7	0.5	0.8	8.2	20.1	0.5	1.0	27.7	20.1	0.5	1.0	27.5
			X	X		20.1	1.0	0.1	1.0	20.3	1.0	0.1	5.1	20.1	1.0	0.1	29.6	20.1	1.0	0.1	29.7
			X		X	20.2	1.0	1.0	1.0	20.4	1.0	1.0	5.3	20.1	1.0	1.0	24.2	20.1	1.0	1.0	24.0

Table 4.8. Results of retrieval of C_{chl} (µg/l), C_{sm} (mg/l), C_{doc} (mg C/l), and H (m). Symbols as in Table 4.7

C_{chl}	C_{sm}			C_{doc}			$H=1$				$H=5$				$H=10$				$H=20$			
	0.1	0.5	1.0	5.0	10.0	15.0	C_{chl}^*	C_{sm}^*	C_{doc}^*	H^*	C_{chl}^*	C_{sm}^*	C_{doc}^*	H^*	C_{chl}^*	C_{sm}^*	C_{doc}^*	H^*	C_{chl}^*	C_{sm}^*	C_{doc}^*	H^*
0.1	X			X			0.1	0.1	5.0	1.0	0.5	0.1	4.9	5.2	0.2	0.1	4.9	24.0	0.2	0.1	4.9	23.8
	X				X		0.1	0.1	10.0	1.0	0.2	0.1	9.9	5.0	0.2	0.1	10.0	5.3	0.2	0.1	9.9	5.3
	X					X	0.1	0.1	15.0	1.0	0.3	0.1	14.7	4.0	0.3	0.1	14.7	4.1	0.3	0.1	14.7	4.1
			X	X			0.1	5.0	0.1	1.0	0.1	5.0	0.1	4.9	0.1	5.0	0.1	6.5	0.1	5.0	0.1	6.5
			X		X		0.1	5.0	5.0	1.0	0.2	5.0	5.0	3.8	0.2	5.0	5.0	3.9	0.2	5.0	5.0	3.9
			X			X	0.2	5.0	15.0	1.0	0.3	5.0	15.0	3.9	0.3	5.0	15.0	3.9	0.3	5.0	15.0	3.9
1.0	X			X			1.0	0.1	5.0	1.0	1.5	0.1	4.8	5.2	1.1	0.1	5.0	23.3	1.1	0.1	5.0	23.1
	X				X		1.0	0.1	10.0	1.0	1.1	0.1	9.9	4.8	1.1	0.1	9.9	5.0	1.1	0.1	9.9	5.0
	X					X	1.0	0.1	15.0	1.0	1.1	0.1	14.7	3.6	1.1	0.1	14.7	3.6	1.1	0.1	14.7	3.6
			X	X			1.0	5.0	0.1	1.0	1.0	5.0	0.1	4.9	1.0	5.0	0.1	6.4	1.0	5.0	0.1	6.4
			X		X		1.1	5.0	5.0	1.0	1.1	5.0	5.0	3.9	1.1	5.0	5.0	3.9	1.1	5.0	5.0	3.9
			X			X	1.1	5.0	15.0	1.0	1.2	5.0	15.0	3.9	1.2	5.0	15.0	3.9	1.2	5.0	15.0	3.9
5.0	X			X			5.0	0.1	5.0	1.0	5.6	0.1	4.9	5.5	5.1	0.1	5.0	22.0	5.1	0.1	5.0	22.0
	X				X		5.1	0.1	10.0	1.0	5.0	0.1	10.0	5.0	5.0	0.1	10.0	5.0	5.0	0.1	10.0	5.0
	X					X	5.3	0.1	15.1	1.0	5.0	0.1	15.0	19.4	5.0	0.1	15.0	19.4	5.0	0.1	15.0	19.4
			X	X			5.1	5.0	0.1	1.0	5.1	5.0	0.1	5.0	5.1	5.0	0.1	19.1	5.1	5.0	0.1	18.5
			X		X		5.1	5.0	5.0	1.0	5.1	5.0	5.0	11.6	5.1	5.0	5.0	12.5	5.1	5.0	5.0	11.7
			X			X	5.2	5.0	15.1	1.0	5.2	5.0	15.0	3.9	5.2	5.0	15.0	3.9	5.2	5.0	15.0	3.9
10.0	X			X			10.1	0.1	5.0	1.0	10.5	0.1	4.9	19.4	10.1	0.1	5.0	19.4	10.1	0.1	5.0	19.4
	X				X		10.2	0.1	10.0	1.0	10.0	0.1	10.0	6.4	10.0	0.1	10.0	6.4	10.0	0.1	10.0	6.4
	X					X	11.2	0.1	15.3	1.0	10.0	0.1	15.0	8.3	10.0	0.1	15.0	8.3	10.0	0.1	15.0	8.3
			X	X			10.1	5.0	0.1	1.0	10.1	5.0	0.1	5.5	10.1	5.0	0.1	18.0	10.1	5.0	0.1	18.2
			X		X		10.2	5.0	5.0	1.0	10.2	5.0	5.0	11.9	10.2	5.0	5.0	11.5	10.2	5.0	5.0	11.5
			X			X	10.4	5.0	15.1	1.0	10.2	5.0	15.0	4.1	10.2	5.0	15.0	4.1	10.2	5.0	15.0	8.4
20.0	X			X			20.3	0.1	5.0	1.0	20.1	0.1	5.0	16.6	20.0	0.1	5.0	15.5	20.0	0.1	5.0	15.5
	X				X		20.7	0.1	10.1	1.0	20.0	0.1	10.0	7.6	20.0	0.1	10.0	7.6	20.0	0.1	10.0	7.6
	X					X	24.3	0.0	15.7	0.9	20.0	0.1	15.0	16.5	20.0	0.1	15.0	16.5	20.0	0.1	15.0	16.5
			X	X			20.2	5.0	0.1	1.0	20.2	5.0	0.1	15.8	20.2	5.0	0.1	14.8	20.2	5.0	0.1	14.9
			X		X		20.3	5.0	5.0	1.0	20.2	5.0	5.0	10.1	20.2	5.0	5.0	10.2	20.2	5.0	5.0	10.2
			X			X	20.7	5.0	15.1	1.0	20.3	5.0	15.0	4.9	20.3	5.0	15.0	4.9	20.3	5.0	15.0	4.9

Table 4.9. Results of application of the two-step scheme including the trans-spectral impacts on retrievals of the concentration vector C and bottom depth H (C_{input} is the CPA concentration used in the model; $C_{ret,I}$, $C_{ret,II}$ are retrieval results obtained with the Levenberg–Marquardt algorithm in the first and second steps)

	C_{chl}, µg/l	C_{sm}, mg/l	C_{doc}, mg C/l	H, (m)
C_{input}	2.00	0.50	0.50	1.00
$C_{ret,I}$	1.13	0.51	0.58	1.00
$C_{ret,II}$	2.20	0.52	0.49	0.99
C_{input}	2.00	0.50	2.0	1.00
$C_{ret,I}$	3.48	1.05	2.66	0.88
$C_{ret,II}$	1.72	0.49	1.99	1.02
C_{input}	2.00	0.50	10.0	1.00
$C_{ret,I}$	2.41	0.15	4.38	1.63
$C_{ret,II}$	2.33	0.44	6.16	0.82
C_{input}	2.00	0.50	0.50	5.00
$C_{ret,I}$	1.31	0.56	0.60	4.78
$C_{ret,II}$	2.21	0.51	0.49	4.96
C_{input}	2.00	0.50	2.0	5.00
$C_{ret,I}$	0.53	4.54	4.07	4.41
$C_{ret,II}$	8.87	0.40	1.18	7.79
C_{input}	2.00	0.50	10.0	5.00
$C_{ret,I}$	1.40	3.07	1.96	8.84
$C_{ret,II}$	1.26	0.41	5.35	29.6

will drop here and henceforth in this section (unless strictly necessary) the wavelength and angular dependencies. Rearranging eq. (4.32) yields:

$$E_u(-0) = Q[L_u(+0) - f_1 E_{sky} - f_2 E_{sun}]/T_{surf}. \qquad (4.33)$$

The downwelling irradiance $E_d(-0)$ just beneath the air–water interface originating from both the diffuse and direct radiation is given by:

$$E_d(-0) = f_3 E_{sky} + f_4 E_{sun}, \qquad (4.34)$$

where f_3 and f_4 are, respectively, the fraction of E_{sky} and E_{sun} that are transmitted through the air–water interface downwards.

In this case the volume reflection, $R(-0)$, which by definition is the ratio $E_u(-0)/E_d(-0)$, can be expressed as follows:

$$R(-0) = \frac{Q_u[(+0) - f_1 E_{sky} - f_2 E_{sun}]}{T_{surf}[f_3 E_{sky} + f_4 E_{sun}]}. \qquad (4.35)$$

Thus, eq. (4.35) yields for the upwelling radiance just above the surface:

$$L_u(+0) = \frac{R(-0)T_{surf}[f_3 E_{sky} + f_4 E_{sun}]}{Q} + f_1 E_{sky} + f_2 E_{sun}. \tag{4.36}$$

As can be seen, the first term on the right-hand side of eq. (4.36) defines the radiance from beneath the water–air interface in the direction of the remote sensor.

Within the framework of the problem, which we discuss here, this light signal should qualify as '*legitimate*' signal. The other two terms are due to the reflection of incident sun and sky radiation (L_{surf}) at the air–water interface into the direction of the remote sensor. According to the notation adopted now, we rewrite eq. (4.36):

$$L_{u,tot}(+0) = L_u(+0) + L_{surf}. \tag{4.37}$$

To quantify the fraction of the 'legitimate' radiance signal we introduce the ratio, which, of course, also implicitly refers to the remote sensor's viewing angle θ_v:

$$P = \frac{L_w(+0)}{L_w(+0) + L_{surf}}, \tag{4.38}$$

Thus, by varying θ_0, wind speed, W, as well as C, it is possible to quantitatively assess the variations in L_{surf} relative to the total water-leaving radiance. This also provides the possibility of identifying the optimal conditions for performing remote sensing of natural waters.

Returning to eq. (4.36), the coefficient f_1 can be expressed in terms of the coefficient of reflection of incident diffuse radiation by the water surface, ρ_{sky}:

$$f_1 = \alpha \rho_{sky} + \beta_1, \tag{4.39}$$

where α is the spectrally dependent ratio of downwelling zenith sky radiance to downwelling sky irradiance, i.e. $L_{sky}(0°)/E_{sky}$; β_1 is the ratio of the upwelling radiance entering the field-of-view of the remote sensor (as a result of sky irradiance being reflected by surface waves) to the incident sky irradiance, E_{sky}.

From Maul (1985), the radiance reflected from a water surface due to wave action ($f_2 E_{sun}$) can be expressed as:

$$f_2 E_{sun} = \frac{\rho(\Omega) E_{sun} \exp[-tg^2\delta/S^2]}{4\pi S^2 \cos\theta_v \cos^4\delta}, \tag{4.40}$$

and hence:

$$f_2 = \frac{\rho(\Omega) \exp[-tg^2\delta/S^2]}{4\pi S^2 \cos\theta_v \cos^4\delta}, \tag{4.41}$$

where Ω = angle of incidence resulting in E_{sun} being reflected into the aperture of the remote sensor; $\rho(\Omega)$ = Fresnel reflectivity for incident solid angle Ω; δ = wave slope resulting in sun glint captured by the remote sensor; and S^2 = mean square wave slope, which is dependent on wind speed W (Cox and Munk, 1954):

$$S^2 = 0.003 + (0.512 \times 10^{-2} W). \tag{4.42}$$

The fraction f_3 in eq. (4.34) may be expressed as:

$$f_3 = 1 - (\rho_{sky} + \beta_3),\qquad(4.43)$$

where ρ_{sky}, as above, is the reflectivity of a flat water surface for a uniformly diffuse sky irradiance; and β_3 is the fraction of incident sky irradiance that is reflected due to wave action.

Analogously, f_4 in eq. (4.34) can be written as:

$$f_4 = 1 - [\rho_{sun}(\theta_0) + \beta_4],\qquad(4.44)$$

where $\rho_{sun}(\theta_0)$, as above, is the reflectivity of a flat water surface when illuminated by direct solar irradiance under a solar zenith angle θ_0; and β_4 is the fraction of incident direct solar irradiance that is reflected due to surface roughness.

The term $[\rho(\theta_0) + \beta_4]$ can be expanded in a series $\rho_w(\theta_0)$ of the form (Cox and Munk, 1954):

$$\rho_w(\theta_0) = [\rho(\theta_0) + \beta_4] = \rho(\theta_0)\{\tfrac{1}{2}[1 + I(k)] + \tfrac{1}{2}[\pi^{-1/2}aS\exp(-k^2)]$$

$$+ \tfrac{1}{4}[bS^2(1 + I(k)) - 2\pi^{-1/2}k\exp(-k^2)]\qquad(4.45)$$

$$+ \tfrac{1}{4}[cS^2(1 + I(k))] + \cdots\}$$

where $k = (2S)^{-1}\operatorname{ctg}\theta_0$;

$$I(k) = 2\pi^{-1/2}\int_0^k \exp(-t^2)\,dt;$$

$$a = \frac{1}{\Im}\frac{d\Im}{d\theta_0};$$

$$b = \frac{1}{2} + \frac{1}{2\Im}\frac{d^2\Im}{d\theta_0};$$

$$c = \frac{1}{2} + \frac{\operatorname{ctg}\theta_0}{2F}\frac{d\Im}{d\theta_0};$$

$$\Im = \rho(\theta_0)\cos\theta_0.$$

Thus, taking into account eq. (4.45),

$$f_4 = 1 - \rho_w(\theta_0),\qquad(4.46)$$

eq. (4.35) can finally be written as:

$$R(-0) = \frac{Q\left[L_u(+0) - f_1 E_{sky} - \dfrac{\rho(\Omega)\exp(-\operatorname{tg}^2\delta/S^2)E_{sun}}{4\pi S^2\cos\psi\cos^4\delta}\right]}{T_{surf}\{[1 - \rho_{sky} - \beta_3]E_{sky} + [1 - \rho_w(\theta_0)]E_{sun}\}}.\qquad(4.47)$$

Given the incident sky and sun irradiances, as well as the fractions f_1, f_2, and the volume reflectance $R(-0)$, it is possible to quantitatively assess $L_u(\lambda, +0)$, and hence solve the task to compare $L_u(+0)$ and L_{surf} entering the field-of-view of the remote sensor. This comparison has to take into account diurnal variations of solar zenith angle, near-surface wind speed, and varying concentration vector C(CPAs).

The following list of input parameters is required for it: incident irradiances E_{sky} and E_{sun}; the sun zenith angle-dependent ratio $F(\theta_0) = E_{sky}/(E_{sky} + E_{sun}(\theta_0))$; the transmission of nadir radiance through the water–air interface T_{surf}; $Q = E_u(-0)/L_u(-0)$, reflecting the angular distribution of the upwelling radiance just beneath the water–air interface; fractions $f_1,\ldots,f_4$; surface reflectivities $\rho_{sky}, \rho_{sun}(\theta_0)$; solar zenith angle θ_0 near-water surface wind speed W; subsurface volume reflectance $R(-0)$; and finally a hydro-optical model, i.e. tabulated spectral values of absorption and backscattering cross-sections, a_i^*, b_{bj}^* of the major CPAs.

For wind speeds in the range 0–10 m, the value of β_1 in eq. (4.39) can be considered as zero (Gordon, 1969), so that

$$f_1 \approx \alpha\rho_{sky}. \tag{4.48}$$

α, being a spectrally dependent ratio of downwelling zenith sky radiance to hemispherical incident sky irradiance, $\alpha = L_{sky}(0^\circ)/E_{sky}$, can be taken, in a first approximation, equal to $1/\pi$ assuming an isotropic angular distribution of the sky radiation under cloudless conditions. At small sun zenith angles and the relative refraction coefficient $n = 1.341$, $\rho_{sky}(0^\circ) \cong 0.0212$ (Jerlov, 1976), and the fraction f_1 can be expressed as

$$f_1 \approx 0.012\alpha. \tag{4.49}$$

During day-time, remote sensing in the temperate climate zone is generally conducted at sun zenith angles from 35° to 65°. This corresponds to values of Ω falling into the range 20°–33°, if the near-surface wind speed is ≤ 5 m/s (Gordon, 1969). With these assumptions, the values f_2 are in the range 8×10^{-8} to 7×10^{-4}. It implies that beyond a sun glint area, the fraction f_2 can be neglected.

There are quantitative assessments that β_3 in eq. (4.43) acquires small negative values, e.g. -0.010 at $W \approx 4$ m/s (Payne, 1972), and -0.014 at $W \approx 7\,7$ m/s (Cox and Munk, 1954). From Jerlov (1976), $\rho_{sky} \approx 0.066$. Consequently, under most conditions of conducting remote sensing of natural waters, $0.93 \leq f_3 \leq 0.95$.

Numerical simulations of $L_u(+0)$ were performed (Pozdnyakov, Lyaskovsky, 2001) for $0 < C_{chl} < 10$ µg/l, $0 < C_{sm} < 5$ mg/l, $0 < C_{doc} < 5$ mg C/l, $0 < W < 15$ m/s, and viewing angles 0°, 10°, 20°, 30°, 40°, and 60°. According to Duntley (1974), the value of T_{surf} is close to 0.544 for a relative index of refraction $n = 1.341$. $\rho_w(\theta_0)$ was first calculated from eq. (4.45) and then substituted into eq. (4.46). $R(-0)$ was obtained from eqs (1.18) and (1.19). The spectral distribution of the diffuse and direct sun incident radiation, as well as the value of the fraction $F(F = E_{sky}/(E_{sky} + E_{sun}(\theta_0))$ were calculated using the models suggested by Gregg and Carder (1990) and Baker and Smith (1997). The fraction F_w was obtained from eq. (1.28). All numerical experiments were conducted for $\phi_v = 0$. Q values were taken from (Loisel and Morel, 2001) for the appropriate hydro-optical conditions.

We first analyse the spectral variations in $L_u(-0)$, because after transmission through the water–air interface, the contribution of this signal to the ratio P (eq.

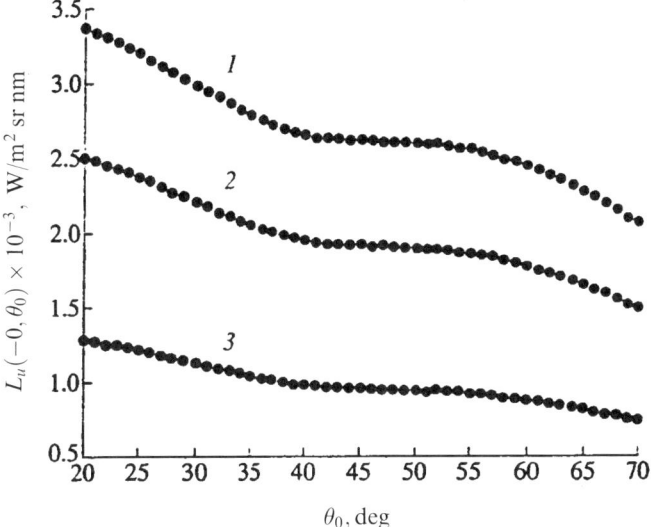

Figure 4.6. Variations of $L_u(-0, \theta_0)$ when the sun zenith angle θ_0 varies between $20°$ and $70°$, for three wavelengths (in nm) of incident light: 560 (*1*), 500 (*2*), and 430 (*3*), for conditions in Lake Ladoga and no wind (see also the text).

(4.38)) will be subject to wind action. Combining eqs (4.32) and (4.37) yields the desired quantity:

$$L_u(-0) = \frac{R(-0)\lfloor(1 - \rho_{sky}\beta_3)E_{sky} + (1 - \rho_w(\theta_0))E_{sun}\rfloor}{Q}. \qquad (4.50)$$

Fig. 4.6 illustrates the calculated variations in $L_u(-0)$ in the direction $\theta_v = 0$ for a variety of solar zenith angles θ_0 several wavelengths λ, calm surface conditions, and the concentration vector typical of the southern region of Lake Ladoga: $C_{chl} = 5.6\ \mu g/l$; $C_{sm} = 1.0\ mg/l$; $C_{doc} = 8.5\ mg/l$ (Kondratyev *et al.*, 1999). As can be seen, with increasing θ_0; radiance $L_u(-0)$ decreases at all wavelengths. An inflection in the dependence of $L_u(-0)$ on θ_0 is distinctly visible at $\theta_0 \approx 40°$. For the simulated hydro-optical situation, namely a considerable amount of *chl* and *doc* in the water column, which is assumed vertically homogeneous, changes in the spectral distribution of $L_u(-0)$, not shown here, are quite large: a drop in the upwelling radiance in the blue and blue-green parts of the visible spectrum accompanied by some increase in the green and green to yellow spectral regions if compared to pure water. This is due to strong absorption of *chl* and *doc* in the short-wave spectrum and enhanced scattering by the hydrosol.

Fig. 4.6 also indicates that with increasing θ_0 the major losses in the upwelling radiance signal occur in the blue. This is mostly due to attenuation of short-wave radiation as the sun approaches the horizon. This example underlines once more that in optically turbid/strongly absorbing waters upwelling radiance is stronger at longer

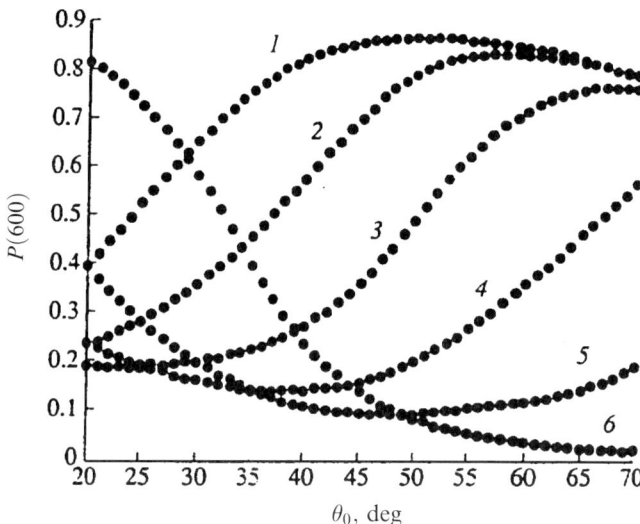

Figure 4.7. Variability of the ratio P at $\lambda = 520$ nm as a function of sun zenith angle θ_0 for the viewing angles $\theta_v : 0°(1)$, $10°(2)$, $20°(3)$, $30°(4)$, $40°(5)$ and $60°(6)$. Near-surface wind speed $W = 5$ m/s.

wavelengths. From this point of view these wavelengths should preferably be exploited in remote sensing.

The variations in the ratio $P(\lambda = 600$ nm) with θ_0 are illustrated in Fig. 4.7 for the following hydro-optical situation: $C_{chl} = 10$ µg/l; $C_{sm} = 10$ mg/l; $C_{doc} = 2$ mg/l. As can be seen, these variations for a wind speed of 5 m/s exhibit a highly complex dependence upon viewing angle θ_v: at $\theta_v \le 30°$, $P(600)$ increases with θ_0, yet at $\theta_v \ge 40°$ the nature of the dependence changes drastically and the ratio $P(600)$ drops with increasing θ_0.

For nadir view ($\theta_v = 0$), curve 1 in Fig. 4.7 shows that although at ($\lambda = 600$ nm) the radiance signal $L_u(-0)$ drops by more than a factor 2 when θ_0 changes from 20° to 70° (see Fig. 4.6), there is still a growth of $P(600)$ with θ_0. This is due to the fact that at a surface wind speed $W = 5$ m/s, the number of facets of the surface waves orientated in such a way that the reflected ray enters the remote sensor's field-of-view decreases with increasing θ_0.

A further increase in viewing angle θ_v results in a substantial change in the ratio $P(600)$ dependence on θ_0 (Fig. 4.7, curve 4 for $\theta_v = 30°$), and at $\theta_v \approx 40°$ the dependence reverses: $P(600)$ now decreases with increasing θ_0 (Fig. 4.7, curves 5, 6). This implies that the aforementioned decrease of $L_u(-0.600)$ with increasing zenith angle of the sun is faster than the one by L_{surf}. This, in turn, indicates an increase in the number of facets oriented in such a way that the reflected ray enters the remote sensor's field-of-view when the latter views the water surface at an angle equal to or in excess of 40°. It should be underlined that this interpretation is based on the probability distribution of slopes of wave facets at a given wind speed, as taken from the Cox and Munk model (their Fig. 8, Cox and Munk, 1954).

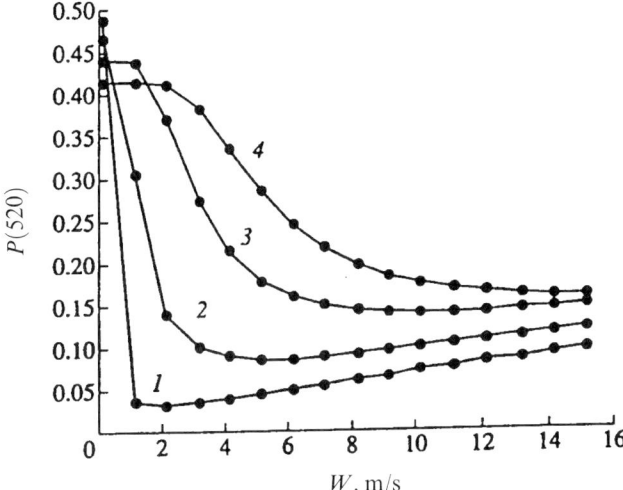

Figure 4.8. Dependence of the ratio P at $\lambda = 520\,\text{nm}$ on wind speed for several sun zenith angles θ_0: $10°(1)$, $20°(2)$, $30°(3)$, $40°(4)$.

Fig. 4.8 displays the results of the numerical experiments aimed at revealing the responsiveness of the ratio P at $\lambda = 520\,\text{nm}$ to wind speed and zenith angle of the sun for nadir view of the air–water interface in the case of clear oligotrophic waters with $C_{chl} = 0.17\,\mu\text{g/l}$; $C_{sm} = 0$; $C_{doc} = 0$.

A wind speed enhancement results in a rapid deterioration of the useful/legit-imate signal at small sun zenith angles θ_0: already a transition from calm conditions to $W \sim 1\,\text{m/s}$ brings about an order of magnitude decrease of $P(520)$. However, with increasing θ_0 the rate of deterioration of $P(520)$ with W becomes less accentuated.

Importantly, when W is as high as $15\,\text{m/s}$, the ratio of $P(520)$ exhibits a certain tendency to stabilization, and even to some recuperation, provided the sun is suffi-ciently high. As a result, the ratio $P(520)$ varies but within rather close limits: from 0.1 to 0.15. This can be explained based on the probability distribution of an elementary facet ξ inclination that ensures that rays reflected from a wind-roughened surface enter the field-of-view of the remote sensor that views the water surface in the nadir direction. As seen from Fig. 4.9, the probability ξ (calculated from eq. (3.12), Bukata *et al.*, 1995) for $\theta_0 = 10°$ is high at low wind speeds ($W \leq 2\,\text{m/s}$), yet it rapidly decreases with enhancing wind speed. At higher values of the sun zenith angle θ_0, the probability ξ first increases with W, but then drops or remains stable at $\theta_0 = 40°$.

Therefore, a straightforward explanation can be provided for the behaviour of the ratio P in Fig. 4.7: at the beginning, the weight of the fraction L_{surf} in P increases with enhancing wind speed W, thus the ratio P decreases. However, a further strengthening of wind results in a lower weight of the fraction L_{surf} in P, and even a certain stabilization of this parameter at a level 0.1–0.15.

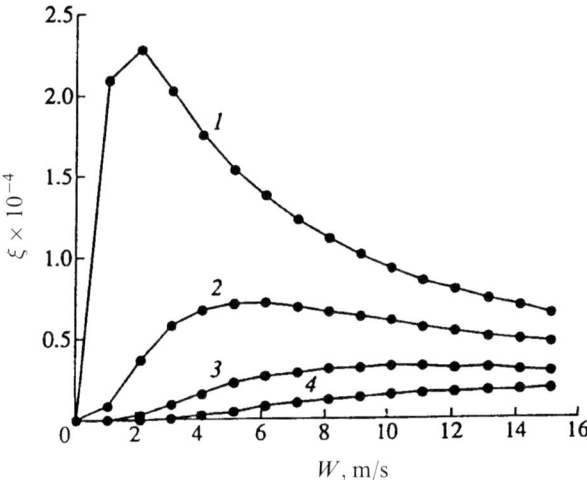

Figure 4.9. Dependence of the probability ξ on near surface wind speed W for $\theta_0 = 10°(1)$, $20°(2)$, $30°(3)$, $40°(4)$.

Thus, the optimal conditions for performing remote sensing of natural waters can be specified as follows: at solar zenith angles of $40°$ (mid-day hours at temperate latitudes) and viewing angles not in excess of $\sim 5°$, at near surface wind speed $W \leq 2$ m/s. With increasing viewing angle θ_v to $\sim 40°$, the fraction of the useful/ 'legitimate' signal declines with its peak value falling in the range of solar zenith angles $65–70°$. As it was shown with the sun close to the horizon, the absolute values of the useful/legitimate signal $L_u(+0)$ drop by about 1.7 to 2.8 times relative to small sun zenith angles (Fig. 4.6).

It should be underlined, however, that the Cox and Munk empirical model of facet slopes was used throughout the presented numerical simulations. Notwith-standing its wide use, this empirical model is not free from methodical shortcomings. In particular, when statistically analysing signals originating from sun glint, the contribution of the radiation reflected from the respective areas of the sky has not been taken into consideration. This, undoubtedly, increases the error, especially at large sun zenith angles. In addition, the orientation of wave slopes at a certain wind direction is neglected in the Cox and Munk model.

Also, other models of the wind-roughened water surface were suggested, e.g. a modified Cox–Munk model (Wozniak, 1999), an analytical model (Longuet-Higgins, 1962), and a model based on direct observations (Martsinkevitch, 1970). Using calculations of the angular structure of spectral radiation reflected from a wind-roughened water surface (Moullamaa, 1964), a comparison of the above mentioned models has shown (Shifrin and Zolotov, 2000) that the value of $L_u(+0)$ may drastically vary (up to a factor of 5) depending on the chosen model. Conse-quently, the establishment of a truly adequate model of the roughened water surface is urgently needed, not only from the perspective of fundamental studies, but also for contemporary remote sensing of the world oceans.

Concluding this section, it should be underlined that due to its relative computational simplicity, the parametric approach used in the simulations discussed above can be used to *operationally* correct the impact of wind action on the useful signal.

5

Retrieval of water quality distributions from SeaWiFS images over non-Case I waters

Spectacular progress has been made since the early 1950s in designing spectrometers/radiometers for remote studies of the Earth's natural environments from aircraft and satellites (Durand *et al.*, 1998; Cracknell *et al.*, 2001). This stimulated national and international programmes that rely on airborne and spaceborne scanners delivering surface reflection spectra from which a wealth of qualitative/quantitative data on the land/water surface properties can be derived. For these deliverables we need procedures capable of removing the impact of the intervening atmosphere on the remotely measured spectrometric data in the visible and near-infrared regions of the spectrum. This is especially true for water bodies where the subsurface upwelling irradiance contributes but a small fraction of the total signal.

Two basic investigations have laid the foundation for remote sensing of open marine and oceanic waters. First, Clarke *et al.* (1970) showed that the algae content in surface waters could be confidently deduced by retrieving the algae chlorophyll concentration from the spectral distribution of the light emerging from beneath the water surface (i.e. from the water colour). Actually, these studies have shown that the ratio of radiances leaving the water surface at 460 and 540 nm can be related to the chlorophyll concentration present in the upper layers of clear ocean waters. Shortly afterwards, Curran (1972) suggested that the atmospheric contribution to the light reflected by a water column and captured by a satellite sensor could be assessed if in addition to 460 nm and 540 nm, the upwelling radiance is also measured at a wavelength >540 nm in order to simultaneously retrieve the aerosol optical thickness over the given location.

This concept was improved substantially by Gordon (1978) who showed that the reflectance around 750 nm could be used to determine the atmospheric impact on the at-satellite radiance also at shorter wavelengths because at $\lambda \geq 750$ nm the ocean water is no longer transparent.

This work became a real milestone and proved to be the beginning of a new era in remote sensing of the world oceans. Since that time, the atmospheric correction studies applied to ocean imageries developed into a specific branch of research, involving solar radiation transfer in scattering and absorbing media, atmospheric physics (with a specific emphasis on microphysics and the optical properties of aerosol particles), and hydro-optics, and optics, of the air–water interface.

A close inspection of the atmospheric correction problem reveals that the ways it might be tackled are essentially dependent on the nature of waters under surveillance. In this chapter we attempt to briefly overview the present state of the art in the domain of removing the atmospheric effects from satellite imagery of non-Case I waters.

Based on this overview, we will substantiate our choice of the atmospheric correction technique for the practical processing of some SeaWiFS images taken over non-Case I waters. This brief overview will then be followed by a section describing the retrieval of water quality parameters related to local and regional hydrodynamic and biological processes.

5.1   ATMOSPHERIC CORRECTION OVER NON-CASE I WATERS[1]

As discussed in earlier chapters, one of the specific features of open ocean waters, usually referred to as Case I waters following the classification by Morel (Morel and Prieur, 1977), is the dominance of phytoplankton and its degradation products for water deviations from pure ocean water colour. These waters are nearly devoid of dissolved organic matter (*doc*) and suspended minerals (*sm*). Being closely correlated with the phytoplankton chlorophyll concentration (*chl*), the degradation products account but for a small fraction of *chl*. Therefore, in such waters, variations in the spectral features of solar light emerging from beneath the water surface can be related confidently to variations in *chl*, even if taking into account some marginal amounts of *doc* released by the living phytoplankton (see also Kondratyev *et al.*, 1990).

Since the chlorophyllous pigments contained in algae cells absorb strongly from 430 to 450 nm and less strongly from 640 to 660 nm (see Chapter 2), the major optical impact of *chl* on the water-leaving radiance is thus confined to the blue portion of the spectrum, leaving the green to red spectral region nearly intact (this is strictly true if *chl* concentration $\leq 0.25\,\mu g/l$ (Gordon and Clark, 1981)). The longer wavelength limit of the water-leaving radiance in the case of open ocean waters is determined by the optical properties of water molecules per se: being strong absorbers at wavelengths $\geq 700\,nm$, they largely prevent light from leaving the water column at such wavelengths.

The basic concept underlying the atmospheric procedures for Case I waters (see a review by Pozdnyakov *et al.*, 2000a) assumes, in conformity with the above, that

---

[1]   This section draws extensively on work done jointly with Stephan Bakan, Max-Planck Institute for Meteorology, Hamburg, Germany (Pozdnyakov *et al.*, 2000a,b).

water-leaving radiance is negligible at wavelengths exceeding 700 nm. Inasmuch as the mean *chl* concentration in the world oceans is assessed at a value as low as $\sim 0.19\,\mu\text{g}/\text{l}$ (O'Reilly *et al.*, 1998), the precision requirements for retrievals of the *chl* concentration are obviously very stringent. This, in turn, entails high demands on the accuracy of atmospheric correction, especially in the blue region of the spectrum. Indeed, owing to the strong absorption band of the phytoplankton chlorophyll from 430 to 450 nm, and the relative insensitivity of the water-leaving radiance at 520–560 nm to the presence of chlorophyll, if $chl < 0.25\,\mu\text{g}/\text{l}$, the upwelling radiance in the blue is the principal information-bearing quantity reflecting the phytoplankton abundance in Case I waters.

Over open ocean areas, the atmosphere is often less turbid with an aerosol optical thickness τ_a ($\tau_a = \int c(z)\,\mathrm{d}z$, z = vertical coordinate, c = beam attenuation coefficient) in the visible not exceeding a value of 0.5 (not infrequently, it lies even under 0.1 (Smirnov *et al.*, 1995)). In addition, the atmospheric aerosol over mid-oceanic areas is assumed to be purely scattering as it consists mostly of sulfate and sea salt particles that are non-absorbing in the visible.

This assumption becomes invalid, however, in the cases of strong advection of continental air masses loaded with either carbonaceous particles or aeolian dust containing ferrous oxides (Kondratyev and Pozdnyakov, 1990a; Kusmierczyk-Michulec *et al.*, 2001). Small values of τ_a and the purely scattering nature of atmospheric aerosols over open oceans significantly simplify the task of atmospheric correction of ocean colour images, because under such conditions the solution of the radiative transfer equation can be reached through well-known and good approximations. This has made atmospheric correction for Case I waters already operational.

Recent studies indicate, however, that the above concept of 'black water' in the infrared is challenged by such factors as air bubbles (see Chapter 2) and water surface roughness (see Chapter 4). According to Yan *et al.* (2002), the top of the atmosphere reflectance deviation that is due to the contribution of oceanic air bubbles (when their concentration is not less than $10^7\,\text{m}^{-3}$) is comparable with the water-leaving reflectance. This air-bubble-driven deviation increases with increasing bubble concentration and coating thickness (see Chapter 2), as well as decreasing aerosol depth τ_a. When $\tau_a \leq 0.3$ at 865 nm, bubble number concentration reaches $10^7\,\text{m}^{-3}$, and thickness of bubble coating $\geq 0.05\,\mu\text{m}$, this deviation grows from 4% in the near-infrared (NIR) to 7–26% in the visible. Thus the 'black pixel' assumption becomes questionable, and, in addition, the growing deviation for shorter wavelengths will also lead to an extra error of the chlorophyll retrieval.

Investigating the impact of water surface roughness on the Rayleigh scattering component of the total upwelling radiance (L_r) at the top of the ocean–atmosphere system, Wang (2002) has shown that at short wavelengths the error in L_r arising from neglect of water roughness (wind speed is 1.9 m/s) is $\leq 1\%$ for solar (θ_0) and sensor viewing zenith angles $\leq 60°$. However, it increases significantly for larger solar and/or sensor zenith angles as well as for stronger near-surface winds. For $\theta_0 \approx 70°$, the error is ≥ 2 or 4% for the wind speeds equal to 7.5 m s^{-1} or 16.9 m s^{-1}, respectively. On the other hand, ΔL_r in the NIR (e.g. 865 nm) varies in the opposite direction to that in the blue. L_r is slightly underestimated there at high wind speed and large solar and viewing zenith angles. Importantly, the errors in L_r associated with water-roughness are most significant at short wavelengths. This combined effect often leads to negative water-leaving radiances at $\lambda \leq 450$ nm, when applying the atmospheric correction based on the 'black water' paradigm. Thus, wind roughness has to be taken into account when correcting atmospheric influence on satellite images in the visible. At present, SeaDAS routine procedure for atmospheric correction of SeaWiFS images, for example, is supplied with Rayleigh look-up tables with wind speed dependence (Wang, 2000).

Coastal marine, as well as inland waters, generally falling into the category of Case II waters, differ dramatically from those of open marine and mid-oceanic areas (see Chapter 2). First of all, Case II waters generally include not solely chlorophyll-bearing biota and associated degradation products (as well as exudated *doc*) but also a variety of suspended inorganic particulates and allochthonic or autochthonic *doc*. Importantly, these components in Case II waters are generally rather abundant and seldom, if at all, co-varying.

The presence of particulate matter in ample amounts in Case II waters increases the probability of backscattering of long-wave photons (Gordon and Morel, 1983). As a result, the water surface cannot be considered black at $\lambda \sim 700$ nm as it was appropriately assumed for 'ideal' Case I waters. On the other hand, phytoplankton and dissolved organic matter are known to absorb strongly at short wavelengths in the visible (see Chapter 2). Co-existing in Case II waters, these two constituents cause a dramatic drop in the blue region of the spectral distribution of upwelling radiance. Therefore, the water surface becomes black at the short wavelengths in highly productive, i.e. meso/eutrophic waters. Nearly all the above constituents of Case II waters absorb light in extensive parts of the visible spectral region, although with different efficiency. Their co-existence in significant amounts results, not infrequently, in very low values of the spectral water reflectance, not only in the blue but also in the green region of the spectrum (Kondratyev *et al.*, 1999).

Unlike over open ocean waters, where the atmosphere is generally remarkably clear, it is typical of coastal marine and notably inland waters that the atmosphere is much more turbid and, apart from purely scattering aerosols, also contain strongly absorbing particulates whose concentration can be quite high. The proximity of Case II waters to natural/anthropogenic sources of aerosol production results in aerosol optical thickness sometimes as high as 0.6–0.7 in the short wavelength region of the visible spectrum (Bartenyeva *et al.*, 1991).

In conformity with the above and the data collected for atmospheres over land by Bartenyeva *et al.* (1991), Skouratov (1997), and some other workers, it can be stated that:

- the probability of aerosol particles containing strongly absorbing matter is much higher than in atmospheres over the ocean;
- the spectral dependence of τ_a varies strongly (e.g. Fig. 5.1);
- relative humidity strongly affects the size distribution of hydroscopic modes of atmospheric aerosols, thus controlling in part aerosol size distributions;

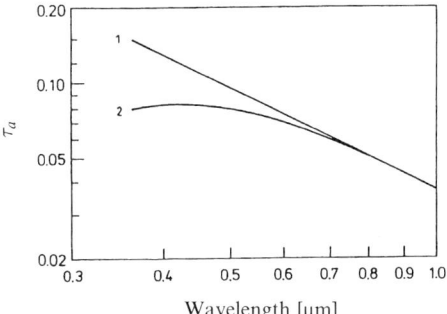

Figure 5.1. Spectral variations of aerosol optical thickness τ_a for a rural area in the St. Petersburg region for different combinations of water vapour pressure (p, atm), water vapour column (W, cm), and relative air mass (m): 1, $p = 6.08$, $W = 0.9$, $m = 2$; 2, $p = 6.08$, $W = 1.5$, $m = 3$.

- τ_a in the blue part of the spectrum can attain values as high as 0.6–0.7, whereas in the red and near-infrared it is mostly several times lower; not infrequently τ_a slightly decreases with λ in the blue part of the spectrum;
- coefficients of correlation between $\tau_a(1000\,\text{nm})$ and $\tau_a(350\,\text{nm})$ can vary between 0.6 and 0.9 and most frequently they lie between 0.7 and 0.8;
- coefficients of correlation between $\tau_a(2000\,\text{nm})$ and $\tau_a(350\,\text{nm})$ are generally significantly lower than between $\tau_a(1000\,\text{nm})$ and $\tau_a(350\,\text{nm})$, rather often they are as low as 0.35, which is an indication of a drastic change in the slope of the spectral distribution of τ_a at $\lambda > 1000\,\text{nm}$.

These data, as we will see below, are important for the further discussion of atmospheric correction techniques for spaceborne imagery over non-Case I waters.

The approaches suggested to performing atmospheric correction in the visible for images taken over non-Case I waters can broadly be categorized as follows: firstly, approaches based on a sophisticated extension of the appropriate Gordon and Morel methods developed for Case I waters, and secondly methods exploiting other methods, such as neural networks and matrix-operator techniques (Schroeder *et al.*, 2002; for a review also see Pozdnyakov *et al.*, 2000b).

It is not the purpose of this book to provide a detailed analysis of atmospheric correction techniques designed so far for processing imageries in the visible (the interested reader is strongly encouraged to capitalize upon the nearly inexhaustible and constantly replenished list of published papers, including Arnone *et al.*, 1998; Moore *et al.*, 1999; Hu *et al.*, 2000a,b,c; Chomko and Gordon, 2001; Antoine and Morel, 1999; Sturm and Zibordi, 2002; Siegel *et al.*, 2000; Wang, 1999a,b; Schroeder *et al.*, 2002, and many others). Instead, we are confining ourselves to a brief description of the method (Ruddick *et al.*, 2000) we used in our applications of the retrieval procedures described in the previous chapters.

The Ruddick *et al.* algorithm, called MUMM, which is an abbreviation of the authors' affiliation: Management Unit of the North Sea Mathematical Models, is an extension of the Gordon and Wang (1994a) standard SeaWiFS atmospheric correction algorithm suggested for open ocean (i.e. Case I) waters.

In MUMM, the assumption of zero water-leaving radiance in the near-infrared (constituting the key feature of the Gordon and Wang approach) is replaced by the assumption that the ratio of aerosol and water-leaving reflectances at 765 and 865 nm are *homogeneous* over the target area.

Following the terminology of Gordon and Wang (1994a), the reflectance at satellite height is defined (dropping for simplicity the spectral dependence notation) as:

$$\rho = \pi L / F_0 \cos \theta_0, \tag{5.1}$$

where F_0 is the extraterrestrial solar flux density, θ_0 is the solar zenith angle, and L is the upward radiance in the viewing direction v of the satellite.

The total at-satellite reflectance ρ_t is a composite of five components:

$$\rho_t = \rho_r + \rho_a + \rho_{ra} + T_v[\rho_w + \rho_{wc}], \tag{5.2}$$

where ρ_r = reflectance resulting from multiple molecular scattering (Rayleigh scattering) in the absence of atmospheric aerosols, ρ_a = reflectance originating from multiple scattering by aerosols in the absence of molecular scattering, ρ_{ra} = reflectance arising from interactions between molecular and aerosol scattering, ρ_{wc} = reflectance of whitecaps, ρ_w = reflectance from the water column, and T_v = diffuse transmittance from the target (pixel of the imagery) to the sensor (satellite).

ρ_r can be computed by use of an exact multiple-scattering code to account for multiple scattering and polarization (look-up tables can be used to speed up the overall atmospheric correction procedure). ρ_{wc} is calculated from Gordon and Wang (1994b) and reduced by a factor of 0.25 (as recommended in the SeaWiFS reprocessing from August 1998).

Thus the Rayleigh-corrected reflectance can be given as:

$$\rho_c = \rho_a + \rho_{ra} + T_v \rho_w. \tag{5.3}$$

To remove the effect of atmospheric oxygen (Wang, 1999b) and (mostly stratospheric) ozone absorption, eq. (5.3) is normalized by dividing it by the two-way ozone and oxygen transmittances:

$$\rho_c' = \rho_a' + \rho_{ra}' + t_{v^*} \rho_w, \tag{5.4}$$

where

$$t_{v^*} = \frac{T_v}{t_{v(oz)} t_{v(O_2)} t_{\theta_0(oz)} t_{\theta_0(O_2)}} = \frac{t_{v(a+r)} t_{\theta_0(a+r)}}{T_0}, \tag{5.5}$$

$T_v = t_{v(a+r)} t_{v(oz)} t_{v(O_2)}$ (atmospheric transmittance under viewing angle v), and $T_{\theta_0} = t_{\theta_0(a+r)} t_{\theta_0(oz)} t_{\theta_0(O_2)}$ (atmospheric transmittance of direct sunlight).

For reasons of simplicity, the prime symbols in eq. (5.4) can be dropped, remembering, however, that henceforth, both ρ_c and its components are normalized by two-way ozone and oxygen transmittances.

Introducing a new notation, called by Ruddick *et al.* (2000) 'the total aerosol scattering reflectance':

$$\rho_{am} = \rho_a + \rho_{ra}, \tag{5.6}$$

and recalling that all terms in eqs (5.2) to (5.6) are wavelength-dependent, eight equations can thus be given for the eight SeaWiFS bands (412, 443, 490, 530, 555, 670, 750, 865 nm) by:

$$\rho_{am}^{(i)} + t_{v^*}^{(i)} \rho_w^{(i)} = \rho_c^{(i)}, \qquad i = 1, \ldots, 8 \tag{5.7}$$

where ρ_c^i is provided from the image data for each pixel, and $t_{v^*}^i$ can be calculated given the viewing geometry and the chosen aerosol model.

With the chosen aerosol model, seven relations between channels 1 to 7 and channel 8 become available:

$$\varepsilon_{s(I)}^{(i,8)} = \frac{g^I[\rho_{am}^{(i)}]}{g^I[\rho_{am}^{(8)}]}, \tag{5.8}$$

where I = number of the aerosol model, subscript s pertains to a single-scattering option $\varepsilon_{s(I)}^{(i,8)}$, and g^I are tabulated (g^I being a coefficient of proportionality between ρ_{as}^i and ρ_{am}^i). Thus the unknowns comprise eight aerosol reflectances including multiple scattering ρ_{am}^{1-8}, eight reflectances from the water body ρ_w^{1-8}, and the best-fit aerosol model index $I = I_0$.

Introduce two assumptions. The ratios:

$$\varepsilon_m^{7,8} \equiv \frac{\rho_{am}^7}{\rho_{am}^8} \quad \text{and} \quad \alpha \equiv \frac{\rho_w^7/T_0^7}{\rho_w^8/T_0^8}$$

are spatially homogeneous within the target sub-scene. α is considered as a calibration parameter to be generally determined for each image. For pure water, the data of Palmer and Williams (1974) give $\alpha = 1.72$.

Combining eq. (5.5) with the second of the above assumptions results in:

$$t_{v^*}^{(7)} \rho_w^{(7)} = \alpha \gamma t_{v^*}^{(8)} \rho_w^{(8)}, \tag{5.9}$$

where

$$\gamma = \frac{t_{v(a+r)}^{(7)} t_{\theta_0(a+r)}^{(7)}}{t_{v(a+r)}^{(8)} t_{\theta_0(a+r)}^{(8)}}. \tag{5.10}$$

Analysing the transmittance tables generated for SeaDAS, Ruddick *et al.* (2000) found that the wavelength variations of γ between 765 and 865 nm are small, lying between 0.98 and 1.00 for typical τ_a and sun and viewing geometries for all coastal aerosol models. Therefore, $\gamma = 1.0$ has been adopted.

With this assumption adopted, it could be written:

$$\rho_c^{(7)} = \varepsilon_m^{(7,8)} \rho_{am}^{(8)} + \alpha t_{v^*}^{(8)} \rho_w^{(8)}, \tag{5.11}$$

$$\rho_c^{(8)} = \rho_{am}^{(8)} + t_{v^*}^{(8)} \rho_w^{(8)}. \tag{5.12}$$

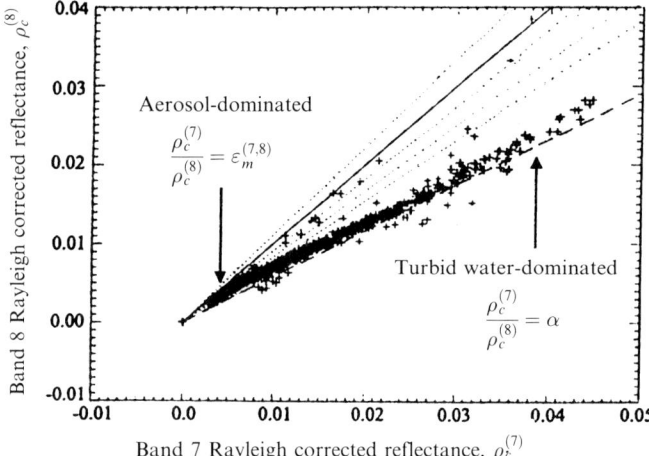

Figure 5.2. Scatterplot of Rayleigh-corrected reflectances at 765 nm and 865 nm for a selected subarea in a SeaWiFS image. The superimposed straight lines correspond to $\rho_c^{(7)}/\rho_c^8 = 1$ (solid line); $= 0.9, 1.10, 1.20, 1.30$ (dotted lines); 1.72 (dashed line).

The solution of eqs (5.11) and (5.12) yields for total aerosol reflectances ρ_{am} and water volume reflectances multiplied by the transmittance t_v:

$$\rho_{am}^{(8)} = \frac{\alpha \rho_c^{(8)} - \rho_c^{(7)}}{\alpha - \varepsilon_m^{(7,8)}}, \tag{5.13}$$

$$t_{v^*}^{(8)} \rho_w^{(8)} = \frac{\rho_c^{(7)} - \varepsilon_m^{(7,8)} \rho_c^{(8)}}{\alpha - \varepsilon_m^{(7,8)}}, \tag{5.14}$$

$$\rho_{am}^{(7)} = \varepsilon_m^{(7.8)} \left[\frac{\alpha \rho_c^{(8)} - \rho_c^{(7)}}{\alpha - \varepsilon_m^{(7,8)}} \right], \tag{5.15}$$

$$t_{v^*}^{(7)} \rho_w^{(8)} = \alpha \left[\frac{\rho_c^{(7)} - \varepsilon_m^{(7,8)}}{\alpha - \varepsilon_m^{(7,8)}} \right]. \tag{5.16}$$

These aerosol reflectances $\rho_{am}^{(7)}$ and $\rho_{am}^{(8)}$ can then be used to find the best-fit aerosol model I_0. This approach, however, implies that the best-fit aerosol model is invariable over the target area, but aerosol reflectance at 865 nm is still allowed to vary freely over the image.

Summarizing the above, the MUMM correction reduces to the following sequence of steps: (a) produce a scatter plot of Rayleigh-corrected reflectances $\rho_c^{(7)}$ and $\rho_c^{(8)}$ for the target area (Fig. 5.2); select the calibration parameters α and $\varepsilon_m^{(7,8)}$ from the scatter plot, (b) deduce $\rho_{am}^{(7)}$ and $\rho_{am}^{(8)}$ from Rayleigh-corrected reflectances $\rho_c^{(7)}$ and $\rho_c^{(8)}$ taking account of non-zero water-leaving reflectances, (c) calculate $\rho_{as(I)}^{(7)}$ and $\rho_{as(I)}^{(8)}$ and then $\varepsilon_{s(I)}^{(7,8)}$ for each candidate aerosol model, (d) select the best two aerosol models through the comparison of $\varepsilon_{s(I)}^{(7,8)}$ with its theoretical counterpart, and determine the interpolation ratio between them, (e) first obtain $\rho_{as}^{(1-6)}$ and then $\rho_{am}^{(1-6)}$ based on the tabulated $\varepsilon_s^{1-6,8}$ for the chosen best-fit aerosol model, (h) subtract $\rho_{am}^{(1-6)}$ from $\rho_c^{(1-6)}$ and divide by the atmospheric transmittance that corresponds to the best-fitting aerosol model to finally yield the desired quantities $\rho_w^{(1-6)}$.

The Ruddick *et al.* algorithm broadly resembles the approach suggested by Arnone *et al.* (1998). The essential difference resides in the number of spectral bands used. Through the inclusion of a third band at 670 nm for atmospheric correction in the Arnone *et al.* algorithm, the constraint requiring spatial homogeneity of the aerosol reflectance ratio can be lifted. It allows allowing both a pixel-by-pixel approach and a treatment of scenes with variable aerosol types. However, the Arnone *et al.* atmospheric correction technique assumes that the subsurface signal in the 670 nm band is not sensitive to phytoplankton chlorophyll on the upwelling radiance, which will often not be the case.

Both algorithms rely upon the Gordon and Wang technique, which selects from look-up tables two bracketing aerosol models from which the spectral dependence of the ratio of Rayleigh-corrected aerosol reflectances known at 765 and 865 nm can be transferred to the remaining SeaWiFS wavelengths. However, as was pointed out above, τ_a, decreasing with λ in the red and near-infrared spectral regions, does not always decline with λ in the short wavelength region (see Fig. 5.1, curve 2), as the SeaDAS aerosol models do. For such instances, errors in atmospheric correction in the blue part of the visible spectrum can be large.

A further source of error in the Ruddick *et al.* approach resides in the assumption of homogeneity of the ratio of aerosol reflectances and water column reflectances at both 765 and 865 nm over the entire target area. This assumption introduces errors for highly heterogeneous fields of atmospheric aerosol type and water composition, respectively, over and within the water area under investigation. The error size depends on the concrete atmospheric and in-water conditions. Therefore, it is difficult to make *a priori* predictions of these errors. Only the practical application of the Ruddick *et al.* approach followed by a thorough validation of the results is capable of justifying, or discarding, the appropriateness of the algorithm.

In our attempts to obtain adequate retrieval of water quality parameters, we compared two available codes of atmospheric correction, viz. the one suggested by Gordon and his team and embedded into the NASA standard SeaWiFS image-processing procedure (SeaDAS), and MUMM. The choice of MUMM was dictated for two reasons: (a) the code of MUMM is freely available and provided in a user-friendly format and (b) the practical application of MUMM to the images taken over non-Case I waters never resulted in negative water-leaving radiances in the short wavelength region. The standard SeaDAS code was tentatively used as the atmospheric correction procedure recommended by NASA as relevant to coastal waters. However, we found that only MUMM proved to be reasonably adequate for, at least, the non-Case I waters studied by us. At the same time the application of the SeaDAS atmospheric procedure turned out to be inadequate for a large number of pixels, and on this basis was rejected.

5.2 DETERMINATION OF BELOW THE WATER SURFACE REFLECTANCE FROM SeaWiFS SPECTRAL DATA

Since atmospheric correction procedures allow us to calculate both downwelling irradiance $E_d(\lambda,+0)$ at the air–water interface and upwelling radiance $L_u(\lambda,+0)$ in the direction of a remote sensor, we can also obtain $R_{rs}(\lambda,+0)$ and from it $R_{rsw}(\lambda,-0)$, which leads to determination of the IOPs via eqs (1.24) and (1.25). Given the spectral distribution of R_{rsw} and eqs (1.24) and (1.25), it is possible to apply the Levenberg–Marquardt multivariate optimization procedure (eqs (4.25)–(4.27)) in order to retrieve the desired CPA concentration vector. This, however, needs an appropriate hydro-optical model (eq. (4.24)).

The quantities $R_{rsw}(\lambda,-0)$ and $R_{rs}(\lambda,+0)$ are related for a flat air–water interface through the following equation (see, for example, Kondratyev et al., 1999):

$$R_{rsw}(\lambda,-0) = R_{rs}(\lambda,+0)\frac{n^2}{(1-\rho_s(\theta_0,F))(1-\rho(\theta_v))}, \qquad (5.17)$$

where $\rho_s(\theta_0,F)$ = coefficient of reflection of incident light (a function of both sun zenith angle θ_0 and the fraction F of diffuse sky radiance: $\rho_s = F(1-\rho_{sun}(\theta_0)) + (1-F)(1-\rho_{sky})$, also see eq. (1.28)); $\rho(\theta_v)$ = coefficient of reflection of the upwelling light at the water–air interface (see eq. (1.22)); n = relative index of refraction ≈ 1.333 (see Chapter 1, eqs (1.22) and (1.23)).

$R_{rs}(\lambda,+0)$ is defined as:

$$R_{rs}(\lambda,+0) = \frac{L_u(\lambda,+0)}{E_d(\lambda,+0)}. \qquad (5.18)$$

$L_w(\lambda,+0)$ and $E_d(\lambda,+0)$ can be approximated by (Wang, 1999a):

$$L_w(\lambda,+0) = [L_w(\lambda,+0)]_N T_{\theta_0}\cos\theta_0, \qquad (5.19)$$

$$E_d(\lambda,+0) = F_0 T_{\theta_0}\cos\theta_0, \qquad (5.20)$$

where, as above, $T_{\theta_0} = t_{\theta_0(a+r)}t_{\theta_0(oz)}t_{\theta_0(O_2)}$ = direct sunlight transmittance for the sun zenith angle θ_0; F_0 = extraterrestrial solar flux at the satellite overflight:

$$F_0 = F_{solar}\left[1 + 0.0167\cos\left(\frac{2\pi}{365}(D-3)\right)\right]^2,$$

F_{solar} = solar constant; $[L_w(\lambda,+0)]_N$ = upwelling radiance emerging from beneath the water surface normalized by the coming solar radiation (available from SeaWiFS data); and D = day of the year.

Combining the above formulae (5.17)–(5.20) yields:

$$R_{rsw}(\lambda,-0) = \frac{[L_w(\lambda,+0)]_N}{F_0(\lambda)}\frac{n^2}{(1-\rho_s(\theta_0,F))(1-\rho(\theta_v))}. \qquad (5.21)$$

Expression (5.21) can equally be written in terms of the water column reflectance ρ_w:

$$R_{rsw}(\lambda, -0) = \frac{\rho_w}{\pi T_{\theta_0}(\lambda)} \frac{n^2}{(1 - \rho_s(\theta_0, F))(1 - \rho(\theta_v))}. \tag{5.22}$$

At the known illumination and viewing geometry and with ρ_w determined from atmospherically corrected SeaWiFS data, $R_{rsw}(\lambda, -0)$ can then be further processed to obtain finally the desired water quality parameters.

5.3 WATER QUALITY PARAMETERS OF ONEGA BAY IN THE WHITE SEA AND THE EASTERN GULF OF FINLAND FROM SeaWiFS IMAGES[2]

In summer 2001, two fairly comprehensive experiments were consecutively carried out in the White Sea (10–15 July) and the eastern Gulf of Finland (20–23 August) under the INCO-COPERNICUS (ICA2-CT-2000-10014 'WHITESEA') and INTAS (INTAS-99-674 'FINGULF') projects, respectively. For several days, overflights of SeaWiFS were accompanied by water sampling just beneath the water surface and *in situ* measurements (water temperature, salinity, electric conductivity, currents, etc.) from on board the Research Vessel *Ecolog* of the Northern Water Problems Institute (NWPI) of the Russian Academy of Sciences in Petrozavodsk, Russia. Water samples were further analysed in the laboratory to yield, among other parameters, the taxonomic composition of phytoplankton and the concentrations of *chl*, *sm*, *doc*, and dissolved oxygen.

In the White Sea the stations were located within Onega Bay (Fig. 5.3). In the eastern Gulf of Finland the sampling stations lay in the area between the Neva River estuary and the 28°E meridian (Fig. 5.4). Typically, the days with SeaWiFS over-flights were not completely cloudless, but the water surface was reasonably smooth in both areas.

5.3.1 Geographical background

5.3.1.1 The White Sea

Located within the subpolar climatic zone between 68° 40′N and 63° 18′N and 32° 00′E and 44° 30′E in the northwestern part of Russia, the White Sea is connected to the Barents Sea, and thus constitutes part of the Arctic Ocean Basin. The water surface area reaches $9 \times 10^4 \, \text{km}^2$, the maximum depth (Fig. 5.5, see colour section) is 350 m, the mean depth is 67 m, the total volume is $6 \times 10^3 \, \text{km}^3$. In terms of the structure–geomorphological classification, the White Sea is a marginal shelf sea. The bottom relief is highly uneven. There are four major bays in the White

---

[2]   For this section we acknowledge contributions by A. V. Lyaskovsky, O. M. Johannessen, L. H. Petterson, L. P. Bobylev, N. N. Filatov, I. A. Neyelov, V. V. Denisov, A. A. Shavykin, and K. S. Khvorostovsky.

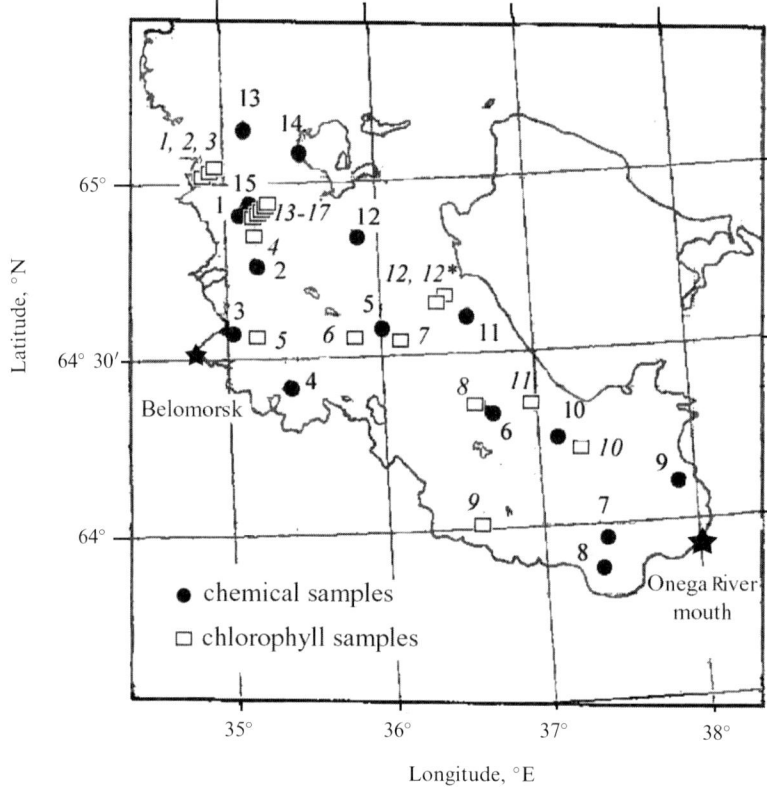

Figure 5.3. Location of validation station in Onega Bay of the White Sea (July, 2001).

Sea: Kandalaksha, Onega, Dvina and Mezen', where the last three are named after the rivers discharging into the sea. The shallow part is located in the northern region of the sea: the major part of the of Mezen' Bay has depths under 20 m. Onega Bay in the west-southern White Sea is even less deep: depths vary between 25 and 5 m. Kandalaksha Bay is the deepest one (except for its apex) (*Hydrometeorology and Hydrochemistry of the Seas*, 1996).

Mean annual salinity in the apices of the bays is about 13–17 per thousand (‰), but in the vicinity of large rivers (North Dvina, Onega, Mezen') it drops to 5–8‰, steadily increasing in the direction towards open parts of the sea, where it surmounts 25‰. At the frontier between the open Barents Sea and the White Sea, the salinity increases to 32‰ (Solyankin *et al.*, 1994).

Several persistent currents largely determine the hydrodynamic patterns of the White Sea (Fig. 5.6). Near to the surface, there is a fairly pronounced current moving from Kandalaksha Bay along the western coast into the Onega Bay coast, and further through the Throat into the Barents Sea. In the Basin there are several circular currents.

Thermohydrodynamic patterns in the White Sea are strongly influenced by lunar

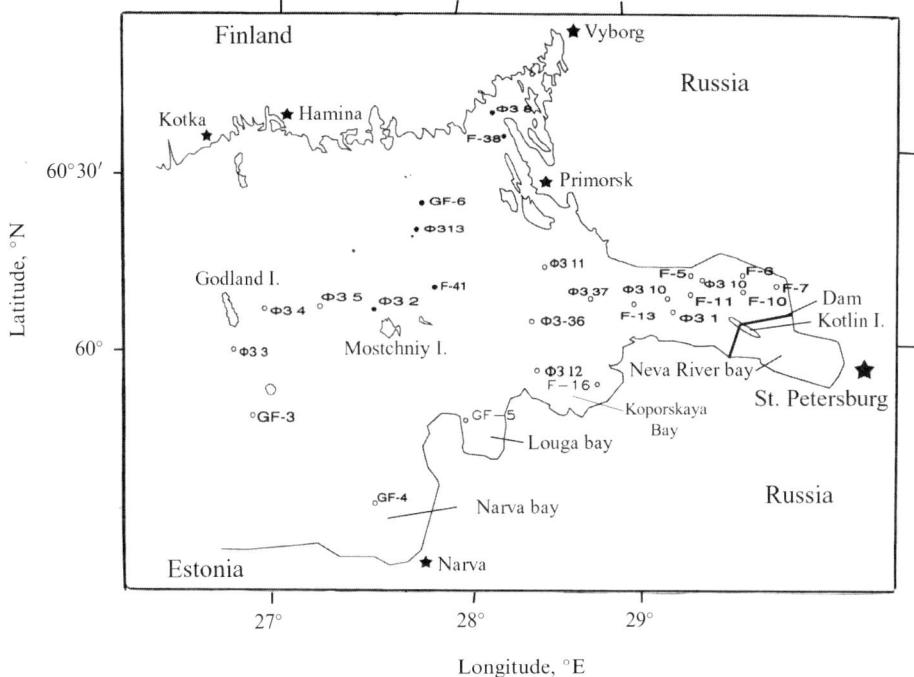

Figure 5.4. Location of validation stations in the eastern Gulf of Finland (August, 2001).

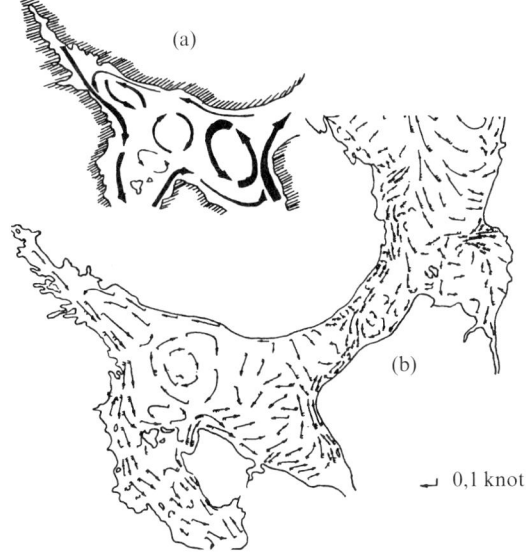

Figure 5.6. General circulation patterns in the White Sea (a) and the quasi-constant currents derived from multi-year observations (b).

tidal waves. One of the manifestations of this phenomenon is a persistent upwelling located at the exit to the Basin from Onega Bay (around the Solovetsky Islands). Tidal movements are at the origin of residual tidal currents, which are formed by nonlinear interactions of tidal currents both with the bottom and with the coastline. The highest tide-driven variations of sea level (about 10 m) are observed in Mezen' Bay. In the Throat, the tidal amplitude decreases to 3.5 m, in the Basin it becomes less than 1 m.

In mid-summer, the water temperature in the White Sea bays, especially in Onega and Dvina Bays, can be as high as 19–20°C in the river deltas, and 10°C at the exit from the bays. In the upwelling region surrounding the Solovetsky Islands the water temperature can be as low as 8–7°C.

The surface layers are mostly supersaturated with respect to dissolved oxygen (in Onega Bay the supersaturation of O_2 (%) reaches up to 110; in Dvina Bay these variations are somewhat broader: up to 125). In the central region, these numbers are more uniform near saturation. The concentrations of total phosphorus (µg-atm/l) are generally within the range 2 to 18; in Onega Bay it varies between 4 and 9; in Dvina Bay between 2 and 8; in Mezen' Bay between 2 and 4; in Kandalaksha Bay between 1 and 12. Very high concentrations of total phosphorus are observed along the northwestern coast of the sea (up to 18), and the eastern cost of Dvina Bay (even up to ~30).

With depth, the dissolved oxygen concentration (%) gradually declines and in near-bottom waters it drops to 75 to 80. Conversely, the total phosphorus concentration (µg-atm/l) increases with depth and reaches ~30 in the deep pelagic waters, remaining at the level of 8–10 and 14–20 at the bottom in Onega and Dvina Bays, respectively.

In summer, the concentration of phytoplankton chlorophyll (µg/l) in the surface waters of Onega and Dvina Bays generally does not exceed 2, whereas in the Basin within the permanent gyre it can reach 4. With the exception only of the Throat, the vertical profile of *chl* peaks in the euphotic zone. In shallow waters (i.e. in Onega, Dvina, and Mezen' Bays) the vertical distribution of *chl* is mostly homogeneous (*Comprehensive Studies of the White Sea Ecosystem*, 1994). Diatoms (such as colonial *Sceletonema*) are the dominant species in the open areas of the sea, whereas, in Onega Bay, Dinophyceae and Chrysophyceae are more abundant. Thus, the waters of the White Sea can be categorized as marine eutrophic waters (Mordasova and Venzel, 1994).

5.3.1.2   The Gulf of Finland

The easternmost part of the Gulf of Finland (60° 30′N, 24°E – 59° 30″N, 30°E) can be qualified as a large estuarine water body into which the Rivers Neva, Kymijoki, and Narva, and several smaller rivers, discharge their waters.

The easternmost part, Neva River Bay, is a shallow water body with high throughflow and a large surface. The characteristic depths here are only 2–3 m and the residence time varies between 2 and 7 days. Intensive wind mixing throughout the entire water column prevents thermal stratification.

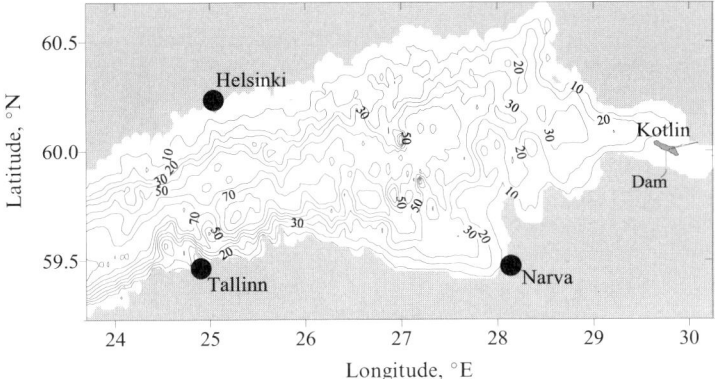

Figure 5.7. Bathymetry of the easternmost part of the Gulf of Finland (numbers of isolines in metres).

A 20-m isobath meanders westwards off Kotlin Island. Depths of 40 to 80 m exist only west of 28°E and they come closer to the coastline of Estonia (Fig. 5.7).

The Gulf of Finland is one of the most polluted areas of the Baltic Sea. In winter and spring, when the easternmost part of the Gulf of Finland is still ice-covered, the nutrient-rich and polluted waters of the River Neva spread westwards under the ice cover.

In summer, owing to specific features of the Neva River estuarine hydro-dynamics, the eutrophication effects due to nutrient-rich waters of the Neva River are largely confined to the estuary. Strong algal blooms can reach more western waters only if the water flow and mixing conditions are favourable enough.

One of the most important features in the phytoplankton species composition is the steadily increasing dominance of blue-green algae (mostly *Oscillatoria* spp.): at present, they predominate in the entire eastern Gulf of Finland. However, in Neva River Bay, separated by the flood-protection dam from the rest of the eastern Gulf of Finland, the water blooms are largely massive developments of cryptomonades. Within Neva River Bay, which is a shallow (the average depth, H, is about 3 m) and nearly freshwater area (salinity ≤0.1‰), the phytoplankton community almost strictly comprises the species indigenous to Lake Ladoga. These waters are less productive than those westward of them owing to very low water transparency (Leppaenen *et al.*, 1997).

Figure 5.8 (Leppaenen *et al.*, 1997) illustrates the measured spatial distribution of chlorophyll concentration in the eastern Gulf of Finland in August, 1995. Clearly, the average concentration of chlorophyll in Neva River Bay exceeds by about two to five times the chlorophyll concentration in the open areas of the eastern Gulf of Finland.

Sparse measurements indicate that there is a relatively weak spatial variability (≤10%) in dissolved oxygen in surface waters. High values of dissolved O_2 are observed in Neva River Bay (8.8–11.4 mg/l). Further off Neva River Bay, the

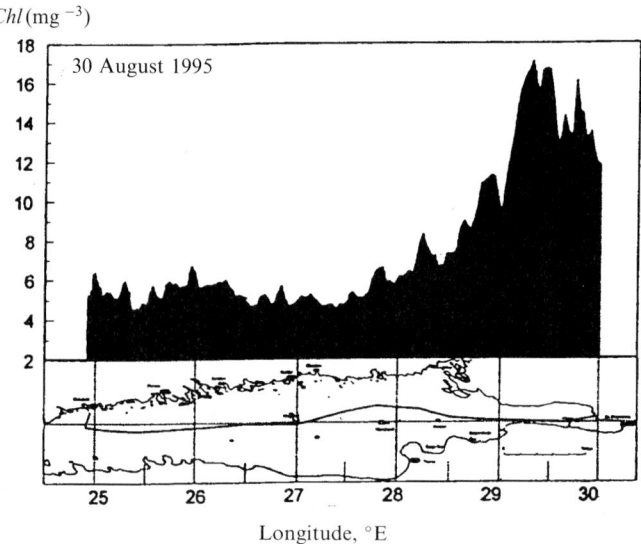

$Chl \, (\mathrm{mg}^{-3})$

Figure 5.8. Chlorophyll-*a* concentration in near surface of the Gulf of Finland in 1995 along the ship-track shown in the lower part.

dissolved oxygen in the surface waters remains high, but steadily declines in near-bottom layers.

There are strong arguments (Leppaenen *et al.*, 1997; Roumantsyev and Drabkova, 1999) that, apart from the estuary and immediately adjacent waters, the ecosystem of the water body in question is nitrogen-limited: $N_{\mathrm{total}} : P_{\mathrm{total}}$ has been decreasing during the last decade, largely due to the phosphorus inflow from bottom sediments as a result of the above mentioned dissolved oxygen deficit in depth.

The Gulf of Finland is hydrodynamically characterized by a two-layer current system with a mean low-saline surface flow from the Neva River estuary towards the west and a compensating deep, saline and phosphorus-rich flow originating from the Baltic Sea. In Neva River Bay, these two layers become mixed entailing a net transport of deep (phosphorus-rich) water to the euphotic zone, whereas in the central and western regions of the Gulf the thermohaline stratification is more common. Therefore, the deep-water phosphorus can reach the euphotic zone only through upwelling or strong turbulent mixing during storms.

Figures 5.9, 5.10, and 5.11 illustrate surface (a) and near-bottom (b) currents, salinity and water temperature in the eastern Gulf of Finland as they are simulated for August mean climate with a three-dimensional circulation model by Neyelov and Oumnov (1997). The model incorporates thermodynamics, and sea ice and is complemented by a biological module developed by Savchuk and Wulf (1996). The model simulates 3-D spatial and temporal distributions of surface, near-bottom and tidal currents, water salinity, water temperature, and ice thickness.

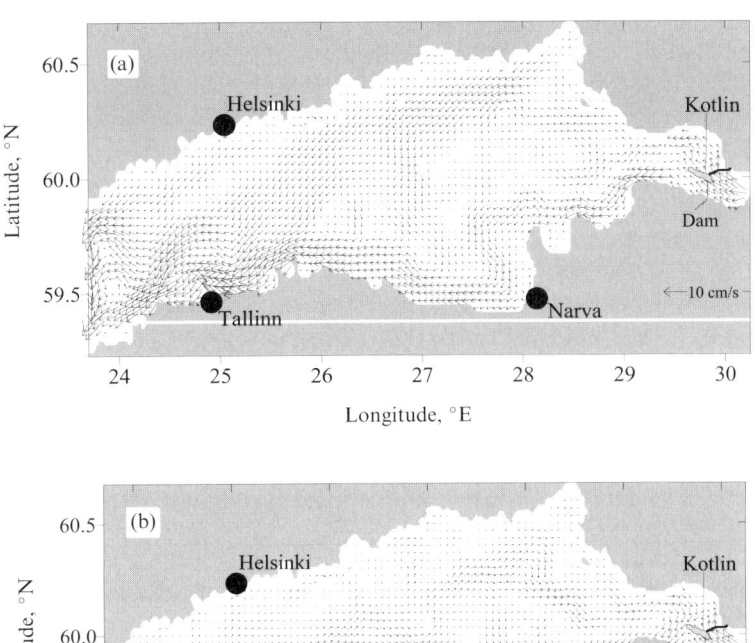

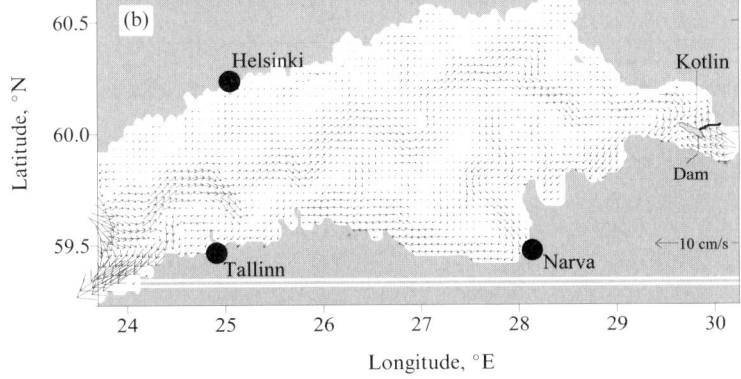

Figure 5.9. Simulated surface (a) and near-bottom (b) currents in the eastern Gulf of Finland during August mean climate conditions. Latitudes in °N, longitudes in °E.

Recapping the essential data given in Sections 5.3.1.1 and 5.3.1.2, it is important to underscore that both water bodies, viz. Onega Bay of the White Sea and the east-ernmost Gulf of Finland, are eutrophic marine environments (Multi-author, 1996; *Hydrometeorology and Hydrochemistry of the Seas*, 1996). Relatively low depths, fairly high summer surface water temperatures, and proximity to runoffs from large rivers rich in nutrients, suspended matter, and soil humus indicate that hydro-optics will correspond to Case II waters with all the ensuing consequences for passive optical remote sensing. This conclusion will further be substantiated by the results of shipborne/laboratory *in situ* measurements, which will be discussed in the rest of this chapter. Besides, the atmospheres over both water bodies are strongly influenced by emissions from local industrial enterprises and air masses coming over from the industrial regions of central Europe. As a result, the atmospheric aerosols there contain appreciable amounts of absorbing particulate matter of

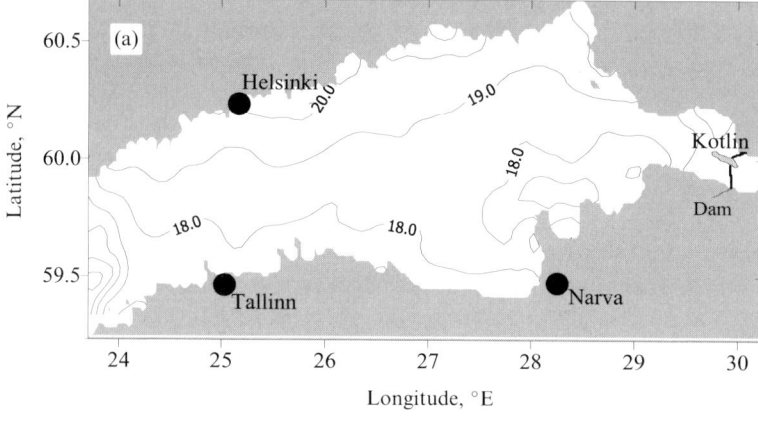

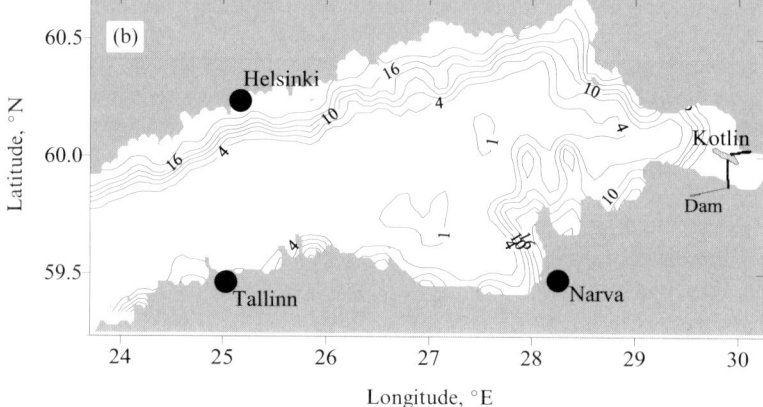

Figure 5.10. Simulated surface (a) and bottom (b) water temperature (°C) in the eastern Gulf of Finland during August mean climate conditions. Latitudes in °N, longitudes in °E.

anthropogenic origin. As was shown in Section 5.1, this is conducive to serious complications for the development of atmospheric correction procedures.

5.3.2   Results of comprehensive field experiments

5.3.2.1   *Onega Bay of the White Sea*

In Onega Bay of the White Sea, the NWPI R/V *Ecolog* has measured basic water parameters and PCAs at 17 stations, whose results are given in Table 5.1. It should be underlined that the *in situ* determinations of concentrations of *chl*, *sm*, and *dom* in both the cases of the White Sea and of the Gulf of Finland (see below) are prone to some inaccuracies arising collectively from instrumentational/methodological errors and errors due to the routinely used sampling procedure performed with the help of a bucket (with the result of a scatter in the series of determinations from one and the

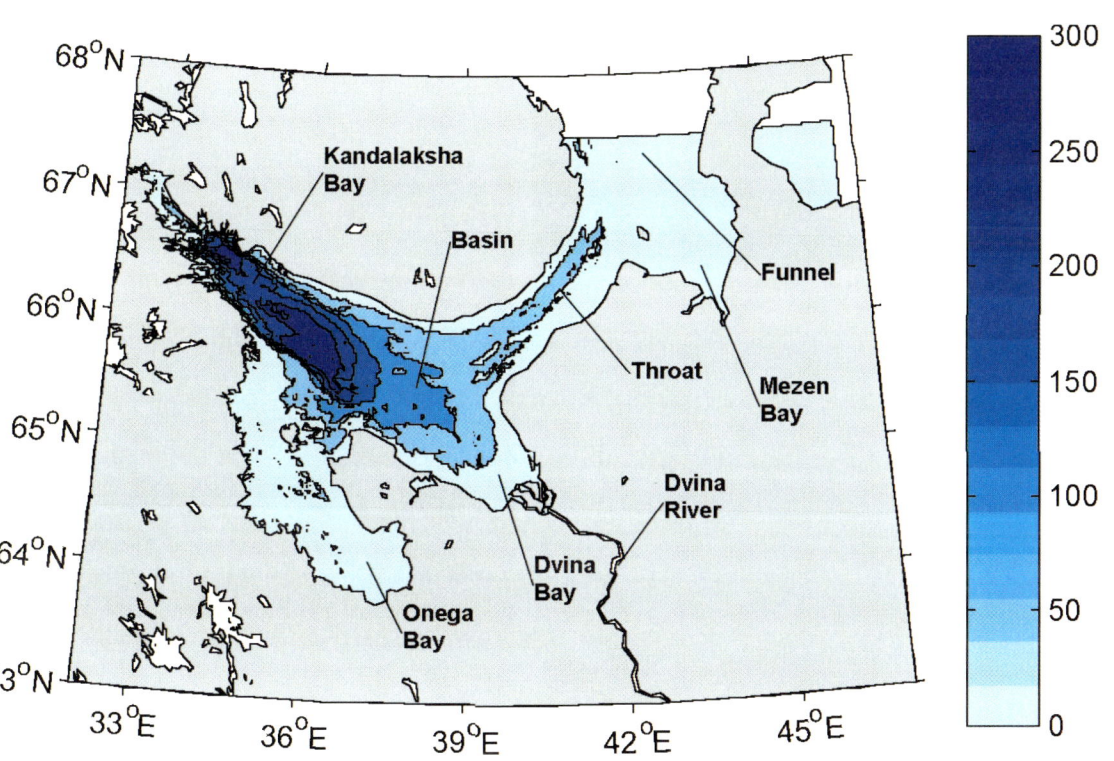

Figure 5.5. Bathymetry of the White Sea. (The scale is in metres.)

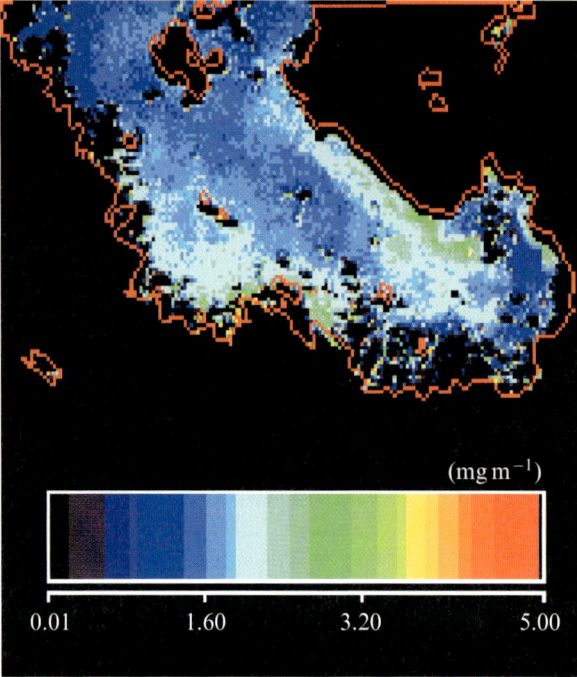

$(\mathrm{mg\,m^{-1}})$

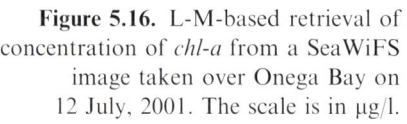

0.01   1.60   3.20   5.00

◄ **Figure 5.15.** SeaDAS-based retrieval of concentration of *chl-a* from a SeaWiFS image taken over Onega Bay on 12 July, 2001. The scale is in µg/l.

▶ **Figure 5.16.** L-M-based retrieval of concentration of *chl-a* from a SeaWiFS image taken over Onega Bay on 12 July, 2001. The scale is in µg/l.

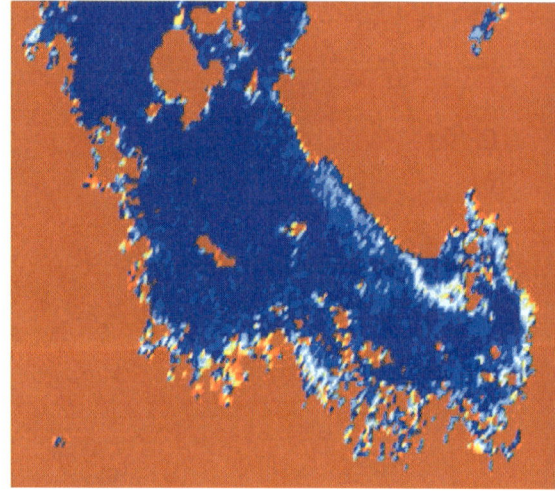

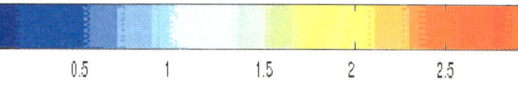

0.5   1   1.5   2   2.5

◄ **Figure 5.17.** L-M-based retrieval of concentration of *sm* from a SeaWiFS image taken over Onega Bay on 12 July, 2001. The scale is in mg/l.

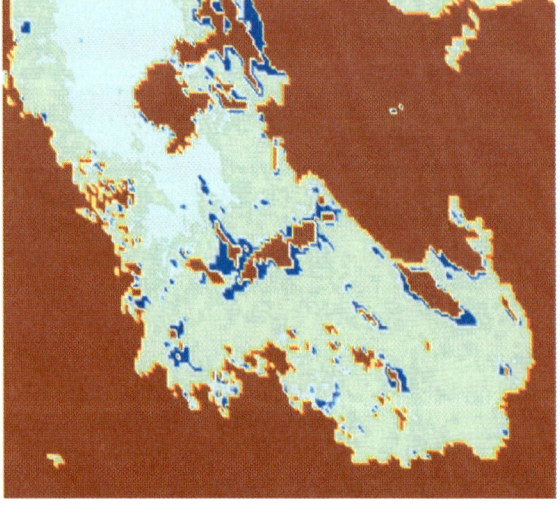

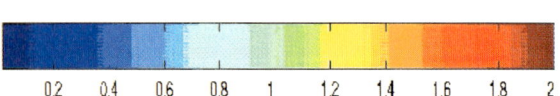

0.2 0.4 0.6 0.8 1 1.2 1.4 1.6 1.8 2

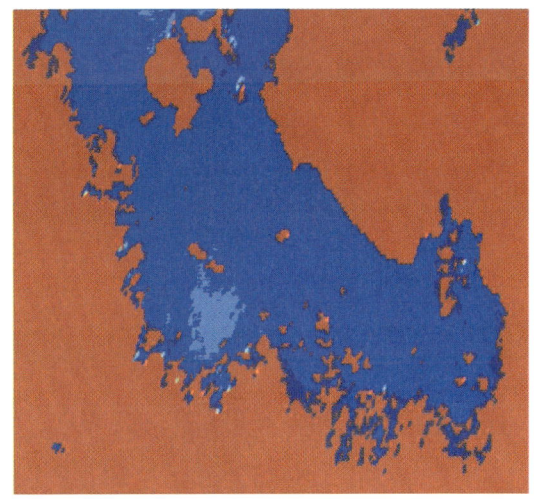

Figure 5.18. L-M-based retrieval of concentration of *doc* from a SeaWiFS image taken over Onega Bay on 12 July, 2001. The scale is in mg C/l.

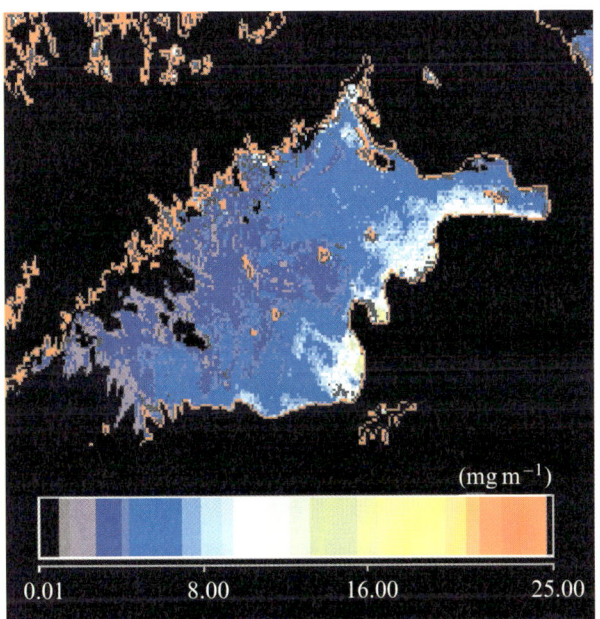

(mg m^{-1})

| 0.01 | 8.00 | 16.00 | 25.00 |

Figure 5.19. SeaDAS-based retrieval of concentration of *chl-a* from a SeaWiFS image taken over the easternmost Gulf of Finland on 20 August, 2001. The scale is in µg/l.

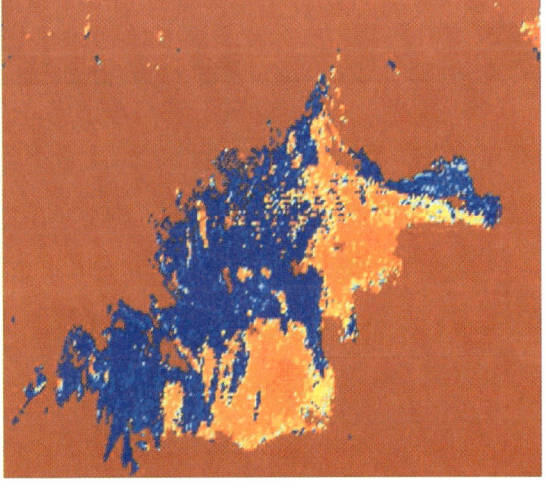

Figure 5.20. L-M-based retrieval of concentration of *chl-a* from a SeaWiFS image taken over the easternmost Gulf of Finland on 20 August, 2001. The scale is in µg/l.

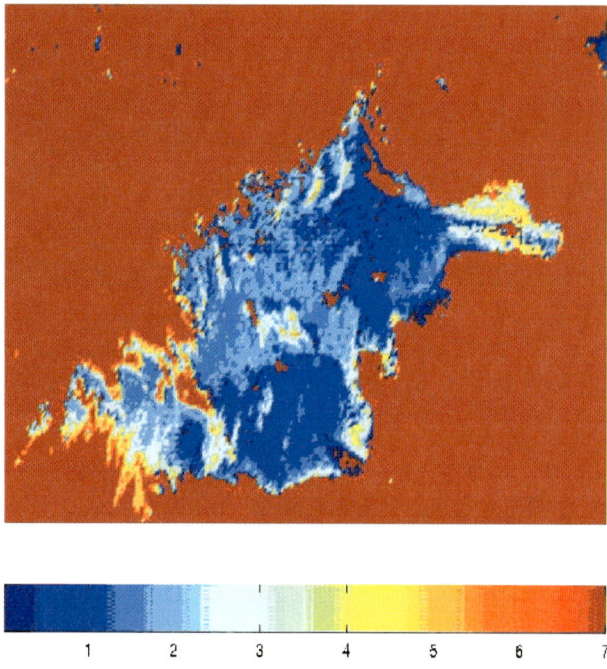

Figure 5.21. L-M-based retrieval of concentration of *sm* from a SeaWiFS image taken over the easternmost Gulf of Finland on 20 August, 2001. The scale is in mg/l.

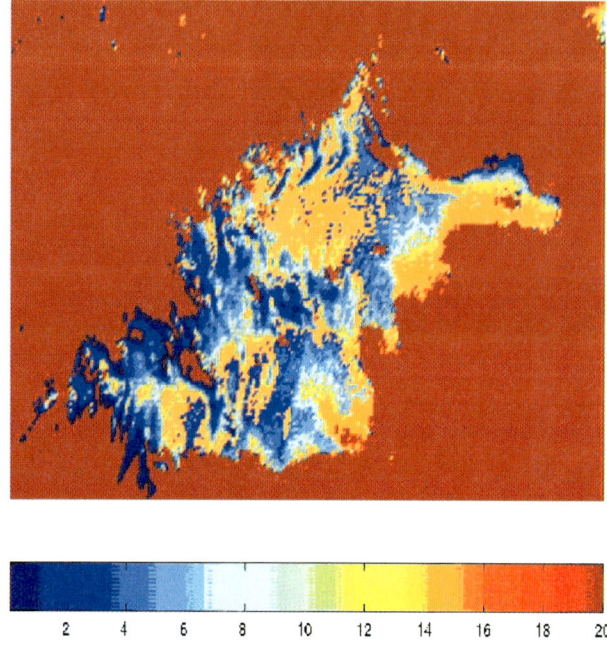

Figure 5.22. L-M-based retrieval of concentration of *doc* from a SeaWiFS image taken over the easternmost Gulf of Finland on 20 August, 2001. The scale is in mg C/l.

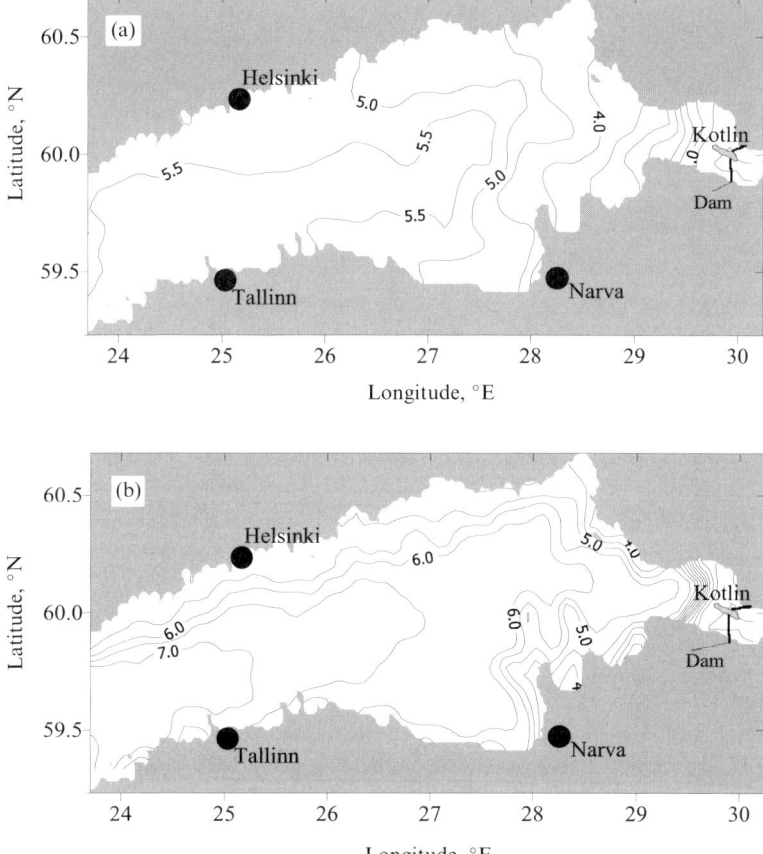

Figure 5.11. Simulated surface (a) and near-bottom (b) water salinity (‰) in the eastern Gulf of Finland for August mean climate conditions.

same bucket given the high degree of inhomogeneity of CPAs in Case II waters). The resultant error is assessed at about 10% for *chl* and *sm* and 5% for *doc* (Petrova, 1990).

Figures 5.12–5.14 illustrate the spatial distributions of water temperature (T_w), salinity (S), and *chl-a* in the surface waters of Onega Bay from *in situ* measurements (both at stations and under way). Clearly, Onega Bay's waters are non-Case I waters.

Figs. 5.12 and 5.13 reveal a specific feature in surface distributions of water temperature T_w and salinity S: there are two 'centres' of enhanced water temperature and low salinity – in the area of the Onega River mouth and near to the small town of Belomorsk, located southward of the Kem' River inlet. From these 'centres', both T_w and S decrease and increase, respectively, along imaginary axes intersecting in the region of the Solovetsky Islands.

The spatial distribution of phytoplankton chlorophyll concentration is fairly uniform over the area of Onega Bay and varies only between about 1 to 2 µg/l

Table 5.1. Results of *in situ* measurements in Onega Bay of the White Sea surface layer, July 2001.

Station number	Date of sampling	*sm*, mg/l	*doc*, mg C/l	*chl*, µg/l fluorimetric method	*chl*, µg/l, spectrometric method
1	10.07	0.25	8.3	—	1.01*
2	10.07	0.25	5.8	0.91	—
3	10.07	0.65	5.9	—	1.12**
4	10.07	0.70	5.8	—	—
5	10.07	0.30	6.3	1.82	—
6	10.07	0.10	5.3	1.45	—
7	10.07	0.35	6.6	0.59	—
8	11.07	0.50	4.9	—	—
9	11.07	0.30	5.9	0.29	—
10	11.07	0.85	5.9	1.21	—
11	11.07	0.70	5.5	1.02	—
12	11.07	0.10	5.3	1.33	—
13	13.07	0.30	5.8	—	—
14	13.07	0.60	5.6	—	—
15	13.07	1.35	5.4	—	1.01*

* Data obtained on 14 July.
** Data obtained at a distance of 10 km from the station location.

(Fig. 5.14), the lower concentrations being found mostly south of the Solovetsky Islands.

As Figs. 5.15–5.18 (see colour section) illustrate, the retrieval results obtained with the use of SeaDAS (Fig. 5.15) and the Levenberg–Marquardt multivariate optimization procedure (L-M) differ strongly although both SeaDAS and L-M retrievals were performed both after atmospherically correcting the SeaWiFS images with the use of the MUMM procedure: utilization of the SeaDAS atmospheric correction code resulted in hopelessly poor restoration of water-leaving radiances in the blue (mostly negative values of $L_w(\lambda, +0)$). It should be explicitly underlined that unlike SeaDAS procedure (see eq. (4.3)), the Levenberg–Marquardt algorithm uses all the SeaWiFS channels.

Table 5.2 compares *in situ* and remotely assessed concentrations of *chl*, *sm*, and *doc* for the L-M code and solely *chl* in the case of application of the SeaDAS code with the MUMM atmospheric correction. The comparison is strongly in favour of the L-M code: the retrieved *chl* concentrations are well within the interval of *chl* values obtained *in situ*, whereas the SeaDAS-produced *chl* values are more than three times higher. In addition, the L-M procedure provides retrievals of *doc* with departures from the respective *in situ* measured values of only 20–25%. However, the retrievals of *sm* show much larger departures from the respective *in situ* values at some stations. Several reasons might lead to the these departures both for *doc* and *sm* (as well as *chl*, although they are much smaller). First of all, *in situ* measurements

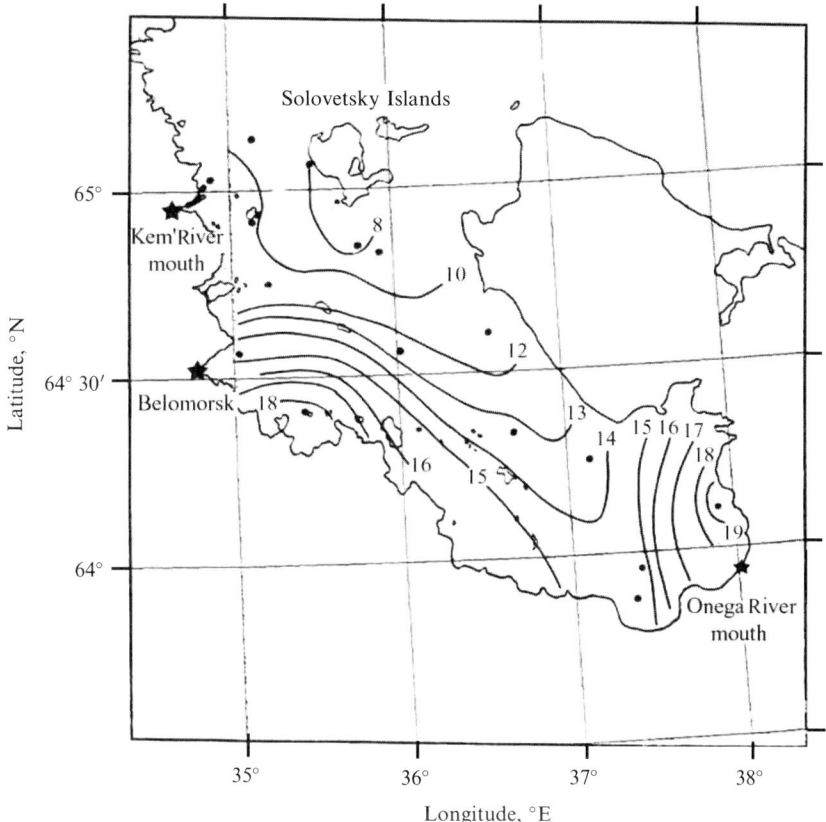

Figure 5.12. Shipborne measurements of near-surface water temperature (°C) in Onega Bay of the White Sea during 10–13 July, 2001. Latitudes in °N, longitudes in °E.

refer to a point on the surface of the bay where the sampling was taken, whereas the SeaWiFS pixel measures $1.1 \times 1.1\,\text{km}^2$, and thus the retrieved value is in reality a spatially averaged value of the desired quantity. Besides, our own experience in *in situ* measurements indicates that the desired quantity can vary by about 20% in a series of samples taken from one and the same pail, when the samples are taken near to the water surface.

Finally, it should be noted that according to Table 5.2, the L-M retrievals of *sm* are invariably higher than their *in situ* counterparts. One of the plausible explanations of this fact is the existence of fine particles in the Onega Bay waters smaller than the filter pores. Consequently, perhaps a considerable amount of *sm* remained unaccounted for.

In addition, the reader should be reminded that we used the hydro-optical model developed not specifically for the White Sea but the one for Lake Ladoga when processing the SeaWiFS images taken over Onega Bay. This definitely will contribute to retrieval errors. However, since Lake Ladoga and the Onega River (as well as North Dina and Mezen' rivers) share very similar watersheds, and are geographically

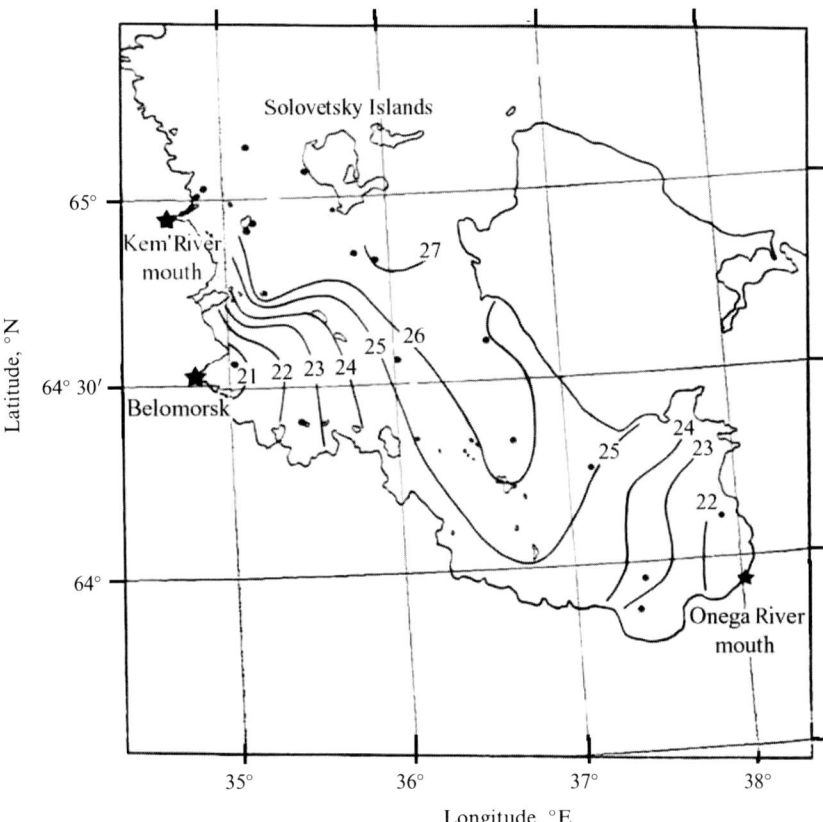

Figure 5.13. Shipborne measurements of surface salinity (‰) in Onega Bay of the White Sea during 10–13 July, 2001. Latitudes in °N, longitudes in °E.

close, we can anticipate (and the above results seem to corroborate this premise) that their hydro-optical models will be very close.

In summary, even given these inaccuracies, the L-M code provides definitely more accurate data than SeaDAS for Onega Bay of the White Sea.

A further analysis of the *chl* field retrieved with SeaDAS from the perspective of the nature of inaccuracies arising from the application of the standard OC4 retrieval algorithm indicates that it gives erroneously high values of *chl* in areas with enhanced concentrations of either *sm* or *doc*. Indeed, according to the water circulation patterns, the waters from the rivers Kem' and Onega move anticlockwise (see Section 5.3.1.1) along the coastline of Onega Bay, being rich in *doc* and *sm*, respectively. It is mainly in coastal areas where OC4 gives too high concentrations of *chl*.

5.3.2.2   The eastern Gulf of Finland

In the easternmost Gulf of Finland, samples of water quality parameters were taken by the NWPI R/V *Ecolog* at 16 stations during 20–23 August, 2001. Their locations

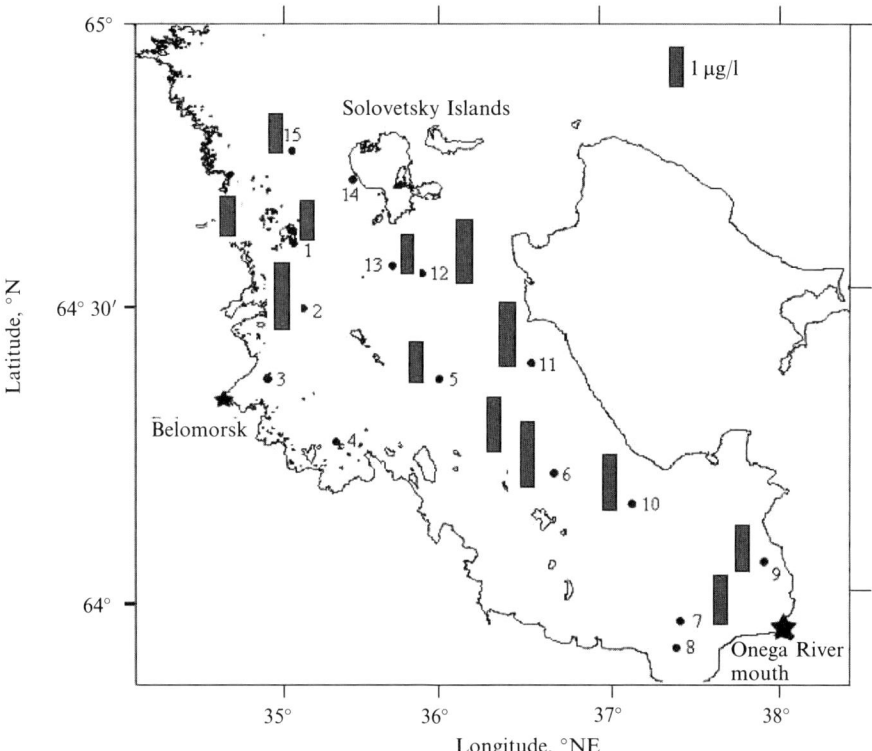

Figure 5.14. Shipborne measurements of *chl* concentration (µg/l) in near-surface water in Onega Bay of the White Sea during 10–13 July, 2001.

Table 5.2. A comparison of water quality variables retrieved for Onega Bay in the White Sea using either SeaDAS or L-M codes, with *in situ* data (July 2001).

Date of *in situ* measurements	*chl-a* (*in situ*), µg/l	*doc* (*in situ*), mg C/l	*sm* (*in situ*), mg/l	Date of SeaWiFS overflight	*chl-a* (SeaDAS), µg/l	*chl-a* (L-M), µg/l	*doc* (L-M), mg C/l	*sm* (L-M), mg/l
10.07.01	1.6	8.3	0.25	10.07.01	3.7	1.3	6.5	0.8
10.07.01	1.1	5.8	0.25	10.07.01	3.0	1.2	5.5	0.7
10.07.01	1.1	5.9	0.65	10.07.01	3.2	1.1	4.5	1.1
10.07.01	1.5	5.8	0.70	10.07.01	4.0	1.5	4.0	1.0
10.07.01	1.8	6.3	0.30	10.07.01	4.2	1.6	4.0	1.0
10.07.01	1.8	5.3	0.10	10.07.01	5.4	1.7	3.9	0.9

(see also Fig. 5.15, in colour section) and the results of these measurements are given in Table 5.3, again clearly indicating a non-Case I water status in this part of the Gulf of Finland.

Unfortunately, there was not a single option of a totally cloudless day during this field experiment. This precluded an exact comparison of retrievals with the *in situ* data for the majority of stations.

Table 5.3. *In situ* measurements of *chl-a*, *sm*, and *doc* concentrations in the Gulf of Finland (August, 2001).

Station	Dates of sampling	*chl-a*, μg/l	*doc*, mg/l	*sm*, mg/l
Φ31	20.08	4.23	10.5	0.8
F-11	20.08	9.00	12.0	1.1
F-10	20.08	17.89	13.5	4.5
F-6	21.08	15.18	14.3	3.1
F-5	21.08	*	*	*
Φ39	21.08	7.32	9.0	0.9
Φ310	21.08	13.12	8.6	0.9
F-13	21.08	9.97	8.3	1.0
F-41	21.08	7.31	8.3	1.2
Φ36	21.08	7.58	7.9	0.8
Φ312	21.08	6.02	9.4	1.0
Φ37	22.08	7.10	7.9	0.7
P1	22.08	11.47	25.5	0.8
20	23.08	13.30	9.0	3.3
32	23.08	25.28	9.0	4.8
15	23.08	11.32	9.0	1.5

Note: All samples have been taken at a 0.5 m depth.
* Sample has not been taken.

It should be acknowledged that, according to Table 5.4, the SeaDAS-based retrieval procedure provides *chl* values sometimes closer to the *in situ* determinations than the L-M algorithm. At the same time, the simultaneous retrievals of *sm* and *doc* by the L-M code compare very well with the *in situ* data. These larger differences in *chl* values might be due to a highly heterogeneous field in this part of the Gulf, resulting in smaller *chl* values found at the stations.

As discussed in the prevous section, comparison of CPA values retrieved from SeaWiFS images (with a spatial resolution of $1.1 \times 1.1 \, km^2$) with *in situ* measurements is not strictly justified, especially given such a limited number of stations and, hence, statistically scarce data (a thorough general discussion of these issues was recently raised by Matthews *et al.*, 2001a,b). Indeed, the sparse *in situ* data can only provide the right *order* of magnitude of the desired quantities. In addition, given the very dynamic nature of fields of *chl*, *sm*, and *doc*, one cannot expect agreement of *in situ* and remote sensing data for different, albeit close, dates.

Figures 5.19 and 5.20 (see colour section) display retrievals using the SeaDAS or L-M code. Comparing the *chl* and *sm* distributions (Figures 5.20 and 5.21, see colour section), it is interesting to point to large areas of low concentrations of *sm* (i.e. transparent waters) that coincide with enhanced concentration of *chl*. In contrast, areas of high concentrations of *sm* correspond to low contents of *chl* just beyond the region of Neva River Bay. These features agree well with the local bathymetry and hydrodynamic patterns discussed above. The same holds for the northern coastline of the Gulf. This example indicates (as does the one related to Onega Bay in the

Table 5.4. *In situ* and retrieved concentrations of *chl-a*, *sm*, and *doc* in the Gulf of Finland in August, 2001.

Date of in situ measurements	chl-a (in situ), μg/l	doc (in situ), mg C/l	sm (in situ), mg/l	Date of SeaWiFS overflight	chl-a (SeaDAS), μg/l	chl-a (L-M), μg/l	doc (L-M), mg C/l	sm (L-M), mg/l
20.08	4.23	10.5	0.8	20.08	10.1	15.0	14.2	0.5
20.08	9.00	12.0	1.1	20.08	8.7	12.0	14.8	2.7
20.08	17.89	13.5	4.5	20.08	8.1	12.1	14.7	3.2
21.08	15.18	14.3	3.1	20.08	8.6	13.5	14.9	1.1
21.08	7.32	9.0	0.9	20.08	9.4	18.6	14.7	1.9
21.08	13.12	8.6	0.9	20.08	7.9	18.0	14.8	2.3
21.08	9.97	8.3	1.0	20.08	7.8	18.0	7.3	1.1
21.08	7.31	8.3	1.2	20.08	6.8	8.1	9.0	1.5
21.08	7.58	7.9	0.8	20.08	8.0	10.0	8.1	1.3
22.08	7.10	7.9	0.7	20.08	7.3	19.0	8.3	1.2
23.08	11.47	25.5	0.8	20.08	5.4	15.0	14.0	1.2

White Sea) the importance of using hydrodynamic and morphometric data when establishing causal relationships between temporal and spatial distributions of water quality parameters.

This is confirmed by the analysis of *doc* distribution in the easternmost part of the Gulf of Finland (Fig. 5.22, see colour section). The *doc* concentrations are most pronounced along the southern coastline owing to topography, water currents, and physicochemistry (Figs 5.7 and 5.9–5.11): the rather shallow, low-salinity areas are subjected to strong currents, land- and river-runoff and waste water effluents rich in humic substances and other dissolved organics. In a similar way, this also might be the case for the area in the vicinity of Helsinki and some other Finnish cities.

A characteristic feature of the SeaWiFS images taken over both Onega Bay in the White Sea and the Gulf of Finland is a pronounced heterogeneity of the spatial distribution of all CPAs – a property inherent to aquatic environments of non-Case I waters. It is the result of the simultaneous influence of neighbouring land areas, bottom topography, hydrodynamics, and biological activity, the latter two changing rapidly with meteorological conditions.

Finalizing this section, it should be underlined that irrespective of the applied retrieval methodology, remote sensing of natural waters in the visible suffers from its vulnerability to cloudiness. This seriously impacts the applicability of this approach when studying *temporal* variations of CPAs. One of the possible ways of overcoming this impediment consists in a synergistic use of sensors operating in different spectral regions, e.g. in the microwave and visible, and, in some cases, also in the infrared.

Outlook

The launch of ERTS-1 (Earth Resources Technology Satellite, later renamed Landsat-1) on 23 July, 1972, brought remote sensing from satellite altitudes to the needs of environmental monitoring. A multispectral scanner (MSS) on Landsat-1 in four solar radiation channels (one green, one red, two near-infrared), was deemed more appropriate for terrestrial than for aquatic monitoring.

Since that time, technical progress in satellite systems for monitoring of coastal and inland water unfolded steadily and with increasing pace. Subsequent Landsat satellites carried improved sensor systems fitted out (Thematic Mapper) with a thermal infrared band and better spectral and spatial resolution. Some sensors were designed specifically for studying the world oceans and water colour (e.g. Seasat and the Coastal Zone Color Scanner (CZCS) aboard Nimbus-7). Technological advancements continued with the development of ocean sensors such as Ocean Colour and Temperature Scanner (OCTS), Polarization and Directionality of Earth's Reflectance (POLDER), Sea-viewing Wide Field-of-view Sensor (SeaWiFS), Moderate Resolution Imaging Spectroradiometer (MODIS), Medium Resolution Imaging Spectroradiometer (MERIS), and the Global Imager (GLI). Further, the forthcoming era of super- and hyper-spectral remote sensing capabilities includes highly sensitive imaging spectroradiometers on scheduled or planned satellites, such as the Australian Resource Information and Environment Satellite (ARIES), Naval Earth Map Observer (NEMO), Land Surface Processes and Interaction Mission (LSPIM), and Spectral Imaging Mission for Science and Application (SIMSA). Details on these and other sensors may be found on the websites of space agencies.

Thus, remote sensing of natural water quality (defined in terms of the combined spectral contributions of aquatic CPAs to remote measurements of water colour) is not restricted by a lack of suitable sensors. Although it can be contested, we see the technology further ahead than remote sensing science. It relates to the entire chain of aquatic remote sensing sequential steps, starting with atmospheric correction and ending in CPA retrieval algorithms. This lag is particularly pronounced for remote sensing of coastal and inland waters, generally ascribed to non-Case I waters.

For remote sensing of non-Case I waters, the following scientific needs can be listed without an attempt to prioritize them.

Further improvement of physical models simulating the interaction of aquatic media with electromagnetic radiation in the visible and near-infrared. Such models should be based on extensive data obtained both *in situ* and in the laboratory. These models have to be supplemented, firstly, by the fluorescence quantum yields of natural fluorophores (such as phytoplankton pigments and dissolved organics), and, secondly, fluorescence band characteristics half-width and the wavelength of maximum emission. But also the fluorescent fraction of *doc* has to be known.

Obviously, the hydro-optical models of the next generation should be specific to water area/water body and vegetation season. We highly recommend a programme on the international level aiming at an atlas of specific inherent properties of CPAs for a variety of water bodies at least for the *main vegetation seasons*. Such an atlas should also include information on bottom types and their spectral albedo, along with the available bathymetric data, predominant currents, average spatial distribution of water temperature and, for marine waters, salinity.

More attention should be given to the relation between colour producing agents (CPAs) at the surface, or in the uppermost centimetres, and the bulk water CPAs. Such relationships should also be basin- and season-specific. They would be very important for the assessment of primary productivity, underwater light climates and sedimentation.

The retrieval algorithms are far from being adequate. Improved ones may be based on new mathematical foundations, although the potential of the L-M multivariate optimization technique, self-organizing neural networks, and matrix-operator methods has not yet been exhausted in the context of passive optical remote sensing. In Chapter 4 we also emphasized the need to assess the impact of surface waves on the upwelling radiance and reflectance just above the surface. It requires a better knowledge of the wave slope distributions, not only for the open ocean but specifically in the coastal zone.

Obviously, comparisons of competing parameter-extraction models/algorithms are required. International workshops should be organized for such intercomparisons, preferably when collecting the results of dedicated international field campaigns. Such a programme has to be based on massively improved ship-based instrumentation, which for applications in Case II water has to be specifically designed (and not 'borrowed' from oceanographers, as often happens at present).

Atmospheric correction is one of the most serious problems in the context of remote sensing of non-Case I waters. Appropriate techniques should adequately account for the atmospheric aerosols, including their absorption properties. Essentially new ideas are needed here since the 'black water' paradigm is either completely inapplicable or results in poor accuracy. Also, cloud elimination procedures should be involved. This, among other things, implies frequent overpasses of satellites with specialized sensors on board.

Special attention should be given to the issue of validation/comparing *in situ* data on concentrations of CPAs and CPA retrieval results (compatibility of spatial scales, the spatial integration problem, statistical analyses) for specific reference sites because of the dynamic nature of aquatic environments. Hence, the calibration and validation of bio-optical and other forward and inverse models must become routine at carefully chosen reference sites.

While equations in physics can sometimes claim precision, biology, biochemistry, and environmental indices are often more governed by statistics. Such statistical analyses, however, depend on sampling schemes and the statistical accuracy of environmental models. In addition, they are site-specific. Hence, remote sensing should not be expected to provide data that are more reliable than those acquired by ground/ship-based monitoring stations. However, remote sensing is the only way to reach a real coverage and thus provide data for ecological models.

Given a fairly frequent areal coverage from remote sensing for atmospheric, aquatic, and surface parameters, ecological models can be not just tested but also used for prediction, their most noble application. We expect attempts to couple general regional circulation models of the atmosphere–ocean/land system with hydro-optical and phytoplankton population modules for algal bloom simulation, their short-term prediction, and biomass production estimates in marginal seas and large inland water bodies.

Passive optical remote sensing of non-Case I waters is a relatively young field of remote sensing. Despite the limitations inherent in water colour remote sensing *per se*, there are many reasons to expect that we will soon see a real breakthrough in this area. However, this will require the concerted action of the international community. We lack not the raw data but a more sophisticated evaluation and subsequent use for modelling and prediction.

References

Aas, E., Højerslev, N. K. (1999) Analysis of underwater radiance observations: apparent optical properties and analytic functions describing the angular distribution. *J. Geophys. Res.*, **104**, 8015–8024.

Aguirre-Gomez, R., Weeks, A. R., Boxall, S. R. (2001) The identification of phytoplankton pigments from absorption spectra. *Int. J. Rem. Sens.*, **22**, No. 2 & 3, 315–338.

Ahn Yu-Hwan, Bricaud, A., Morel, A. (1992) Light backscattering efficiency and related properties of some phytoplankters. *Deep-Sea Res.*, **39**, No. 11/12, 1835–1855.

Allali, K., Bricaud, A., Herve, K. (1997) Spatial variations in the chlorophyll-specific absorption coefficients of phytoplankton and photosynthetically active pigments in the equatorial Pacific. *J. Geophys. Res.*, **102**, No. C6, 12413–12423.

Ammenberg, P., Flink, P., Lindel, T. (2002) Bio-optical modelling combined with remote sensing to assess water quality. *Int. J. Rem. Sens.*, **23**, No. 8, 1621–1638.

Anonymous (1957) Commission Internationale de l'Eclairage (CIE), *Vocabulaire International de l'Eclairage*, CIE Publication No. 1, 2nd edition, Paris, 136 pp.

Antoine, D., Morel, A. (1999) A multiple scattering algorithm for atmospheric correction of remotely sensed ocean colour (MERIS instrument): principle and implementation for atmospheres carrying various aerosols including absorbing ones. *Int. J. Rem. Sens.*, **20**, No. 9, 1875–1916.

Aponasenko, A. D., Shchur, L. A., Filimonov, V. S., Lopatin, V. N. (1997) Structural studies of aquatic ecosystems based on the concept of phase boundaries on the surface of suspended matter. *J. Aquatic Ecology*, **1**, No. 2, 13–25.

Arnone, R. A., Martinolich, P., Gould, Jr, R. W., Stumpf, R., Ladner, S. (1998) Coastal optical properties using SeaWiFS. In: *Proc. Ocean Optics XIV*, Kailua-Kona, Hawaii, USA, November 10–13.

Atkinson, P. M., Tatnall, A. R. T. 1997. Neural networks in remote sensing. *Int. J. Rem. Sens.*, **18**, No. 4, 699–709.

Augusteteijn, M. F., Clemens, L. E., Shaw, K. A. (1995) Performance evaluation of texture measures for ground cover identification in satellite images by means of a neural network classifier. *IEEE Transactions on Geoscience and Remote Sensing*, **33**, 616–626.

Austin, R. W. (1974) The remote sensing of spectral radiance from below the ocean surface. In: *Optical Aspects of Oceanography* (N. G. Jerlov & E. Steeman-Neilsen, Eds). London: Academic Press, pp. 317–344.

Babin, M., Stramski, D. (2002) Light absorption by aquatic particles in the near-infrared spectral region. *Limnol. Oceanogr.*, **47**, No. 3, 911–915.

Babin, M., Morel, A., Gentili, B. (1996) Remote sensing of sea surface Sun-induced chlorophyll fluorescence: consequences of natural variations in the optical characteristics of phytoplankton and the quantum yield of chlorophyll *a* fluorescence. *Int. J. Rem. Sens.*, **12**, 2417–2448.

Bagheri, S., Zetlin, C., Dios, R. (1999) Estimation of optical properties of nearshore water. *Int. J. Rem. Sens.*, **20**, No. 17, 3393–3397.

Baker, K. S., Smith R. C. (1997) Irradiance transmittance through the air–water interface. In: *Ocean Optics X, Proc. SPIE Int. Soc. Opt. Eng.*, **1302**, pp. 556–565.

Barlett, J. S. (1996) The influence of Raman scattering by seawater and fluorescence by phytoplankton on ocean colour. M.S. Thesis (Halifax: Dalhousie University).

Barlett, J. S. (1997) A comparison of models of sea-surface reflectance incorporating Raman scattering by water. In: *Ocean Optics XII, Proc. SPIE Int. Soc. of Opt. Eng.*, **2963**, 592–602.

Barlett, J. S., Voss, K. J., Sathyendranath, S., Vodacek, A. (1998) Raman scattering by pure water and seawater. *Appl. Opt.*, **37**, 3324–3332.

Barnard, A. H., Zaneveld, J. R. V., Pegau, W. S. (1999) *In situ* determination of the remotely sensed reflectance and the absorption coefficient: closure and inversion. *Appl. Opt.*, **38**, No. 24, 5108–5116.

Bartenyeva, O. D., Nikitinskaya, N. I., Sakourov, G. G., Veselova, L. K. (1991) *Atmospheric Columnar Transparency in the Visible and Near Infrared Regions of the Spectrum.* Leningrad: Gidrometeoizdat, 224 pp. (in Russian).

Bertilsson, S., Travnik, L. J. (2000) Photochemical transformation of dissolved organic matter in lakes. *Limnol. Oceanogr.*, **45**, No. 4, 753–762.

Berwald, J., Stramski, D., Mobley, C. D., Kiefer, D. D. (1998) Effect of Raman scattering on the average cosine and diffuse attenuation coefficient of irradiance in the ocean. *Limnol. Oceanogr.*, **43**, 564–576.

Bidigare, R. R., Ondrusek, M. E., Morrow, J. H., Kiefer, D. A. (1990) Bacterioplankton and solar light. In: *Ocean Optics X, Proc. SPIE Int. Soc. Opt. Eng.*, **1302**, 291–302.

Bishop, C. M. (1995) *Neural Networks for Pattern Recognition.* Oxford: Clarendon Press, 482 pp.

Boivin, L. P., Davidson, W. F., Storey, R. S. (1986) Determination of the attenuation coefficient of visible and ultraviolet radiation in heavy water. *Appl. Opt.*, **25**, 877–882.

Bouman, H. A., Platt, T., Sathyendranath, S., Irwin S., Wernard, M. R., Kraay, G. W. (2000) Bio-optical properties of the subtropical North Atlantic. II. Relevance to models of primary production. *Mar. Ecol. Prog. Ser.*, **200**, 19–34.

Bowers, D. G., Harker, G. E. L., Stepan, B. (1996) Absorption spectra of inorganic particles in the Irish Sea and their relevance to remote sensing of chlorophyll. *Int. J. Rem. Sens.*, **17**, No. 12, 2449–2460.

Bowers, D. G., Kvatzer, S. M., Morrison, J. R. (2001) On the calibration and use of in situ ocean colour measurements for monitoring algal blooms. *Int. J. Rem. Sens.*, **22**. No. 2 & 3, 359–368.

Boynton, G. C., Gordon, H. (2000) Irradiance inversion algorithm for estimating the absorption and backscattering coefficients of natural waters: Raman-scattering effects. *Appl. Opt.*, **39**, No. 18, 3012–3022.

Bricaud, A., Morel, A., Prieur, L. (1983) Optical efficiency factors of some phytoplankters. *Limnol. Oceanogr.*, **28**, No. 5, 816–832.

Bricaud, A., Babin, M., Morel, A., Claustre, H. (1995) Variability in the chlorophyll-specific absorption coefficients of natural phytoplankton: analysis and parameterization. *J. Geophys. Res.*, **100**, C7, 13321–13332.

Bricaud, A., Morel, A., Barale, V. (1999) MERIS potential for ocean colour studies in the open ocean. *Int. J. Rem. Sens.*, **20**, No. 9, 1757–1769.

Brivio, P. A., Giardino, C., Zilioli, E. 2001. Determination of chlorophyll concentration changes in Lake Garda using an image-based radiative transfer code for Landsat TM images. *Int. J. Rem. Sens.*, **22**, No. 2 & 3, 487–502.

Buckton, D., O'Mongain, E., Danaher, S. 1999. The use of neural networks for estimation of oceanic constituents based on the MERIS instrument. *Int. J. Rem. Sens.*, **20**, No. 9, 1841–1851.

Buiteveld, H., Hakvoort, J. H. M., Donze, M. (1994) The optical properties of pure water. In: *Ocean Optics XII, Proc. SPIE Int. Soc. Opt. Eng.*, **2258**, 174–183.

Bukata, R. P., Jerome J. H., Bruton, J. E. (1985a) *Application of Direct Measurements of Optical Parameters to the Estimation of Lake Water Quality Indicators.* Environment Canada Inland Waters Directorate Scientific Series, No. 140, 35 pp.

Bukata, R. P., Bruton, J. E., Jerome J. H. (1985b) Particulate concentrations in Lake St. Clair as recorded by a shipborne multispectral optical monitoring system. *Rem. Sens. Environ.*, **25**, 201–229.

Bukata, R. P., Jerome, J. H., Kondratyev, K. Ya., Pozdnyakov, D. V. (1995) *Optical Properties and Remote Sensing of Inland and Coastal Waters.* Boca Raton, FL: CRC Press, 362 pp.

Bukata, R. P., Pozdnyakov, D. V., Jerome, J. H., Tanis, F. J. (2001) Validation of a radiometric color model applicable to optically complex water bodies. *Rem. Sens. Env.*, **77**, 165–172.

Campbell, E. W., Esaias, W. E. (1983) Basis for spectral curvature algorithms in remote sensing of chlorophyll. *Appl. Opt.*, **22**, 1084–1093.

Carder, K. L., Chen, F. R., Lee, Z. P., Hawes, S. K., Kamykowski, D. (1999) Semianalytic moderate-resolution imaging spectrometer algorithms for chlorophyll *a* and absorption with bio-optical domains based on nitrate-depletion temperatures. *J. Geophys. Res.*, **104**, No. C3, 5403–5421.

Chomko, R. M., Gordon, H. R. (2001) Atmospheric correction of ocean color imagery: test of the spectral optimization algorithm with the Sea-viewing Wide Field-of-View Sensor. *Appl. Opt.*, **40**, No. 18, 2973–2984.

Church, M. J., Ducklow, H. W., Korl, D. M. (2002) Multiyear increases in dissolved organic matter inventories at Station ALONA in the Northern Pacific Subtropical Gyre. *Limnol. Oceanogr.*, **47**, No. 1, 1–10.

Ciolli, A. M., Lewis, M. R., Cullen, J. J. (2002) Assessment of the relationships between dominant cell size in natural phytoplankton communities and the spectral shape of the absorption coefficient. *Limnol. Oceanogr.*, **47**, No. 2, 404–417.

Clarke, G. K., Ewing, G. C., Lorenzen, C. J. (1970) Spectra of backscattered light from the sea obtained from aircraft as a measure of chlorophyll concentration. *Science*, **167**, 1119–1121.

Claustre, H., Morel, A., Hooker, S. B. (2002) Is desert dust making oligotrophic waters greener? *Geophys. Res. Letters*, **29**, No. 10, 107-1–107-4.

Coble, P. G. (1994) Investigation of the geochemistry of dissolved organic matter in coastal waters using optical properties. In: *Ocean Optics XII, Proc. SPIE Int. Soc. Opt. Eng.*, **1302**, 291–302.

Coble, P. G., Brophy, M. M. (1996) Investigation of the geochemistry of dissolved organic matter in coastal waters using optical properties. In: *Ocean Optics XII, Proc. SPIE Int. Soc. Opt. Eng.*, **494**, 56–61.

Comprehensive Studies of the White Sea Ecosystem (1994) (V. V. Sapozhnikov, Ed.). Moscow: VNIRO, 289 pp. (in Russian).

Cox, C., Munk, W. (1954) Measurement of the roughness of the sea surface from photographs of the sun's glitter. *J. Opt. Soc. Am.*, **44**, No. 11, 838–849.

Cracknell, A. P., Newcombe, S. K., Black, A. F., Kirby, N. E. (2001) The ABDMAP (Algal Bloom Detection, Monitoring and Prediction) Concerted Action. *Int. J. Rem. Sens.*, **22**, No. 2 & 3, 205–247.

Culver, M. E., Perry, M. J. (1997) Calculation of solar-induced fluorescence in the surface and subsurface waters. *J. Geophys. Res.*, **102**, No. C5, 10 563–10 572.

Curran, R. J. (1972) Ocean color determination through a scattering atmosphere. *Appl. Opt.*, **11**, No. 8, 1857–1866.

Desiderio, R. A. (2000) Application of the Raman scattering coefficient of water to calculations in marine optics. *Appl. Opt.*, **39**, No. 12, 1893–1894.

Dierssen, H. M., Smith, R. C. (2000) Bio-optical properties and remote sensing ocean color algorithms for Antarctic Peninsula waters. *J. Geophys. Res.*, **105**, No. C11, 263013–26312.

DiToro, D. M. (1978) Optics of turbid estuarine waters: approximations and applications. *Water Res.*, **12**, 97–125.

Doerffer, R. (1981) Factor analysis in ocean colour interpretation. In: *Marine Science*, Vol. 13, *Oceanography from Space* (J. F. R. Gower, Ed.). New York: Plenum Press, pp. 339–345.

Doerffer, R. (1992) Imaging spectroscopy for detection of chlorophyll and suspended matter. In: *Imaging Spectroscopy: Fundamentals and Prospective Applications* (F. Toselli & J. Bodechtel, Eds). Brussels and Luxembourg: ECSC, EEC, and EAEC, pp. 215–257.

Doerffer, R., Fischer, J. (1994) Concentrations of chlorophyll, suspended matter, and gelbstoff in Case II waters derived from satellite coastal zone color scanner data with inverse modeling methods. *J. Geophys. Res.*, **99**, No. C4, 7457–7466.

Doubovick, O. V., Ostchepkov, S. L., Lapenock, T. V. (1994) Iteration/regularization method of solution to nonlinear inverse problems and its application to interpreting water leaving radiance spectra. In: *Proc. Russian Academy of Sciences: Ocean and Atmospheric Physics*, **30**, No. 1, 106–113 (in Russian).

Doxaran, D., Froidefond, J.-M., Lavender, S., Castaing, P. (2002) Spectral signature of highly turbid waters: application with SPOT data to quantify suspended particulate matter concentrations. *Rem. Sens. Env.*, **81**, 149–161.

Duntley, S. Q. (1974) *Ocean Color Analysis*. San Diego: Scripps Institution of Oceanography, SIO Ref. 74-10, 67 pp.

Durand, D., Pozdnyakov, D. V., Sandven, S., Cauneau, F., Wald, L., Jacob, A., Kloster, K., Miles, M. (1998) *Characterization of Inland and Coastal Waters with Airborne and Spaceborne Sensors*. Report for the Centre for Earth Observation. Bergen, Norway: NERSC, 179 pp.

Estep, L. (1992) Estimation of bottom reflectance spectra. *Int. J. Rem. Sens.*, **13**, 393–397.

Estep, L. (1994) Bottom influence on the estimation of chlorophyll concentration in water from remotely sensed data. *Int. J. Rem. Sens.*, **15**, No. 1, 205–214.

Fadeev, V. V., Chekaliuk, A. M., Choubarov, V. V. (1982) Nonlinear laser fluorimetry of organic substances. *Proc. Russian Academy of Sciences*, **262**, No. 2, 338–342 (in Russian).

Fahlman, S. E., Lebiere, C. (1990) The cascade-correlation learning architecture. In: *Advances in Neural Information Processing Systems 2* (D. Touretzky, Ed.). San Mateo, CA: Morgan Kaufmann, pp. 524–532.

Fischer, J., Kronfeld, U. (1990) Sun-stimulated chlorophyll fluorescence. 1: Influence of oceanic properties. *Int. J. Rem. Sens.*, **11**, 2125–2147.

Fisher, J., Doerffer, R. (1987) An inverse technique for remote detection of suspended matter, phytoplankton and yellow substances from CZCS measurements. *Adv. Space Res.*, **7**, No. 2, 21–26.

Fournier, G., Forand, J. L. (1994) Analytic phase function for ocean water. In: *Ocean Optics XII, SPIE Int. Soc. Opt. Eng.*, **2258**, 191–201.

Frette, O., Erga, S. R., Stamnes, J. J., Stamnes, K. (2001) Optical remote sensing of waters with vertical structure. *Appl. Opt.*, **40**, No. 9, 1478–1487.

Gallie, E. A., Murtha, P. A. (1992) Specific absorption and backscattering spectra for suspended minerals and chlorophyll *a* in Chilko Lake, British Columbia. *Rem. Sens. Env.*, **39**, 103–118.

Garver, S., A., Siegel, D. A. (1997) Inherent optical property inversion of ocean color spectra and its biogeochemical interpretation. I. Time series from the Sargasso Sea. *J. Geophys. Res.*, **102**, 18607–18625.

Ge, Yu., Voss, K. J., Gordon, H. R. (1995) *In situ* measurements of inelastic light scattering in Monterey Bay using solar Fraunhofer lines. *J. Geoph. Res.*, **100**, No. C7, 13227–13236.

Gege, P. (1998) Characterization of the phytoplankton in Lake Constance for classification by remote sensing. *Arch. Hydrobiol. Spec. Issues: Advanc. Limnol.*, **53**, 179–193.

Gege, P., Heege, T., Albert, A., Thiemann, S. (2001) Determination of water constituents in Lake Constance, Germany. Proceedings of the Workshop on Remote Sensing and Resource Management in Nearshore and Inland Waters, 23.10.2001, Wolfville, Canada, pp. 78–79.

Gohin, F., Druon, J. N., Lampert, L. (2002) A five channel chlorophyll concentration algorithm applied to SeaWiFS data processed by SeaDAS in coastal waters. *Int. J. Rem. Sens.*, **23**, No. 8, 1639–1661.

Gordon, H. R. (1978) Removal of atmospheric effects from satellite imagery of the ocean. *Appl. Opt.*, **17**, No. 10, 1631–1636.

Gordon, H. R. (1979) Diffuse reflectance of the ocean: the theory of its augmentation by chlorophyll *a* fluorescence at 685 nm. *Appl. Opt.*, **18**, 1161–1166.

Gordon, H. R. (1989) Dependence of the diffuse reflectance of natural waters on the sun angle. *Limnol. Oceanogr.*, **34**, 1631–1636.

Gordon, H. R. (1999) Contribution of Raman scattering to water-leaving radiance: a re-examination. *Appl. Opt.*, **38**, No. 15, 3166–3174.

Gordon, H. R., Clark, D. K. (1981) Clear water radiances for atmospheric correction of coastal zone scanner imagery. *Appl. Opt.*, **20**, 4175–4180.

Gordon, H. R., Morel, A. Y. (1983) *Remote Assessment of Ocean Colour for Interpretation of Satellite Imagery*. New York: Springer-Verlag, 115 pp.

Gordon, H. R., Wang, M. (1994a) Retrieval of water-leaving radiance and aerosol optical thickness over the oceans with SeaWiFS: a preliminary algorithm. *Appl. Opt.*, **33**, No. 3, 443–452.

Gordon, H. R., Wang, M. (1994b) Influence of oceanic whitecaps on atmospheric correction of ocean-color sensors. *Appl. Opt.*, **33**, 7754–7763.

Gordon, H. R., Brown, O. B., Jacobs, M. M. (1975) Computed relationships between the inherent and apparent optical properties of a flat homogeneous ocean. *Appl. Opt.*, **14**, 417–427.

Gordon, J. I. (1969) *Directional Radiance (Luminance) of the Sea Surface*. San Diego: Visibility Lab. Publ., SIO Ref. 69-20, 50 pp.

Gorlenko, V. M., Dubinina, G. A., Kuznetsov, S. I. (1974) *Ecology of Aquatic Microorganisms*. Moscow: Nauka Press, 289 pp. (in Russian).

Gould, R. W., Arnone, R. A., Maritinolich, P. M. (1999) Spectral dependence of scattering coefficient in Case 1 and Case 2 waters. *Appl. Opt.*, **38**, No. 12, 2377–2383.

Gould, R. W., Arnone, R. A., Sydor, M. (2001) Absorption, scattering, and remote-sensing reflectance relationships in coastal waters: testing a new inversion algorithm. *J. Coastal Res.*, **17**, No. 2, 328–341.

Gower, J. F. R., Borstad, G. A. (1990) Mapping of phytoplankton by solar-stimulated fluorescence using an imaging spectrometer. *Int. J. Rem. Sens.*, **11**, 313–320.

Gower, J. F. R., Borstad, G. A. (1993) Use of imaging spectroscopy to map solar-stimulated chlorophyll fluorescence, red tides and submerged vegetation. In: *Proc. 16th Canadian Symposium on Remote Sensing Ottawa*, Publication of the Canada Remote Sensing Soc., pp. 95–98.

Gower, J. F. R., Doerffer, R., Borstad, G. A. (1999) Interpretation of the 685 nm peak in water leaving radiance spectra in terms of fluorescence, absorption and scattering, and its observation by MERIS. *Int. J. Rem. Sens.*, **20**, No. 9, 1771–1786.

Green, S., Blough, N. V. (1994) Optical absorption and fluorescence properties of chromophoric dissolved organic matter in natural waters. *Limnol. Oceanogr.*, **39**, 1903–1916.

Gregg, W. W., Carder, K. L. (1990) A simple spectral solar irradiance model for cloudless maritime atmospheres. *Limnol. Oceanogr.*, **35**, 1657–1675.

Gregg, W. W., Chen, F. C., Mezaache, A. L. (1993) The simulated SeaWiFS data set, version 1. In: *SeaWiFS Technical Report Series*, Vol. 9 (S. B. Hooker, Ed.), NASA Tech. Memo. 1999-20692. Greenbelt, MD: Goddard Space Flight Center, pp. 1–17.

Grew, G. W. (1981) Real-time test of MOCS algorithm during Superflux 1980. In: *The Chesapeake Bay Plume Study: Superflux 1980* (J. W. Campbell & J. P. Thomas, Eds). Hampton, VA: Langley Research Center, pp. 121–128.

Gross, L., Thiria, S., Frouin, R. (1999) Applying artificial neural network methodology to ocean colour remote sensing. *Ecol. Modelling*, **120**, 237–246.

Gross, L., Thriria, S., Frouin, R., Mitchel, B. G. (2000) Artificial neural networks for modeling the transfer function between marine reflectance and phytoplankton pigment concentration. *J. Geophys. Res.*, **105**, No. C2, 3483–3495.

Hakvoort, J. H. M. (1994) *Absorption of Light by Surface Water*. The Netherlands: Delft University Press, 145 pp.

Haltrin, V. I., Kattawar, G. W. (1993) Self-consistent solutions to the equation of transfer with elastic and inelastic scattering in ocean optics: 1. Model. *Appl. Opt.*, **32**, 5356–5367.

Heinemann, T., Fischer, J. (1997) Simultaneous retrieval of oceanic and atmospheric properties using satellite remote sensing measurements. *Proc. SPIE Int. Soc. Opt. Eng.* **2963**, 691–698.

Hoge, F. E., Lyon, P. E. (1996) Satellite retrieval of inherent optical properties by linear matrix inversion of oceanic radiance models: an analysis of model and radiance measurement errors. *J. Geophys. Res.*, **101**, No. C7, 16631–16648.

Hoge, F. E., Swift, R. N. (1986) Chlorophyll pigment concentration using spectral curvature algorithms: an evaluation of present and proposed satellite ocean color sensor bands. *Appl. Opt.*, **25**, No. 20, 3677–3682.

Hu, C., Carder, K. L., Muller-Karger, F. E. (2000a) How precise are SeaWiFS ocean colour estimates? Implications of digitization-noise errors. *Rem. Sens. Env.*, **76**, 239–249.

Hu, C., Carder, K. L., Muller-Karger, F. E. (2000b) Atmospheric correction of SeaWiFS imagery: assessment of the use of alternative bands. *Appl. Opt.*, **39**, No. 21, 3573–3581.

Hu, C., Carder, K. L., Muller-Karger, F. E. (2000c) Atmospheric correction of SeaWiFS imagery over turbid coastal waters: a practical method. *Rem. Sens. Env.*, **74**, 195–206.

Hulbert, E. O. (1945) Optics of distilled and natural water. *J. Opt. Soc. Amer.*, **35**, 698–705.

Hulst, H. C. van de (1957) *Light Scattering by Small Particles*. New York: Wiley, 670 pp.

Hutchinson, G. E. (1957) *Treatise on Limnology*. New York: Wiley, 458 pp.

Hydrometeorology and Hydrochemistry of the Seas (1994) Vol. III, *The Baltic Sea*, 2nd edn. St. Petersburg: Gidrometeoizdat, 435 pp. (in Russian).

Hydrometeorology and Hydrochemistry of the Seas (1996) Vol. II, *The White Sea*, 1st edn. St. Petersburg: Gidrometeoizdat, 250 pp. (in Russian).

Jerlov, N. G. (1976) *Marine Optics*, Elsevier Oceanography Series 14. Amsterdam: Elsevier, 213 pp.

Jerome, J. H., Bukata, R. P., Miller, J. R. (1996) Remote sensing reflectance and its relationship to optical properties of natural water. *Int. J. Rem. Sens.*, **17**, No. 1, 43–52.

Jerome, J. H., Bukata R. P., Bruton, J. E. (1998a) Tracking the propagation of solar ultraviolet radiation: dispersal of ultraviolet photons in natural waters. *J. Great Lakes Res.*, **27**, 666–680.

Jerome, J. H., Bukata R. P., Bruton, J. E. (1998b) Utilizing the components of vector irradiance to estimate the scalar irradiance in natural waters. *Appl. Opt.*, **27**, 4012–4018.

Karhu, M., Mitchel, B. (2001) Seasonal and non seasonal variability of satellite-derived chlorophyll and colored dissolved organic matter concentration in the California Current. *J. Geophys. Res.*, **106**, No. C2, 2517–2559.

Karnaukhov, B. N. (1988) *Spectral Analysis of Cells in Prospective of Ecology and Environment Protection (Cellular Biomonitoring)*. Pousthichino, Russia: Scientific Centre for Biological Studies, 125 pp. (in Russian).

Kavzoglu, T., Mather, P. M. (1999) Pruning artificial neural networks: an example using land cover classification of multi-sensor images. *Int. J. Rem. Sens.*, **20**, No. 14, 2787–2803.

Keiner, L. E., Brown, C. W. (1999) Estimating oceanic chlorophyll concentrations with neural networks. *Int. J. Rem. Sens.*, **20**, No. 1, 189–194.

Kirk, J. T .O. (1976) Yellow substance (Gelbstoff) and its contribution to the attenuation of photosynthetically active radiation in some inland and coastal south-eastern Australian waters. *Australian J. Marine Freshwater Res.*, **27**, 61–71.

Kirk, J. T. O. (1981) *Monte Carlo Procedure for Simulating the Penetration of Light into Natural Waters*, CSIRO (Australia) Division of Plant. Tech. Paper No. 36, 16 pp.

Kirk, J. T. O. (1984) Dependence of relationship between inherent and apparent optical properties of water on solar altitude. *Limnol. Oceanogr.*, **29**, 350–356.

Kirk, J. T. O. (1991) Volume scattering function, average cosines, and the underwater light field. *Limnol. Oceanogr.*, **36**, 455–467.

Kirk, J. T. O. (1999) Multiple scattering of a photon flux: implications for the integral average cosine of the underwater light field. *Appl. Opt.*, **38**, No. 15, 3134–3140.

Kondratyev, K. Ya., Pozdnyakov, D. V. (1990a) *Atmospheric Aerosol Models*. Moscow: Nauka, 104 pp. (in Russian).

Kondratyev, K. Ya., Pozdnyakov, D. V. (1990b) *Optical Properties of Natural Waters and Remote Sensing of Phytoplankton*. Leningrad: Nauka Press, 190 pp. (in Russian).

Kondratyev, K. Ya., Moskalenko, N. K., Pozdnyakov, D. V. (1983) *Atmospheric Aerosol.* Leningrad: Gidrometeoizdat, 230 pp. (in Russian).

Kondratyev, K. Ya., Pozdnyakov, D. V. (1996) Ozone depletion-induced UVB radiation impact on aquatic ecosystems. II. Hydrooptical aspects of the problem. *Earth Obs. Remote Sens.*, **6**, 105–114 (in Russian).

Kondratyev, K. Ya., Pozdnyakov, D. V., Isakov, V. Yu. (1990) *Radiation and Hydrooptical Experiments on Lakes.* Leningrad: Nauka Press, 115 pp. (in Russian).

Kondratyev, K. Ya., Filatov, N. N., Johannessen, O. M., Melentyev, V. V., Pozdnyakov, D. V., Ryanzhin, S. V., Shalina, E. V., Tikhomirov, A. I. (1999) *Limnology and Remote Sensing: a Contemporary Approach* (Kirill Ya. Kondratyev & N. N. Filatov, Eds). Chichester, UK: Springer-Praxis, 408 pp.

Krotkov, N., Vasilkov, A. P. (2000) Reduction of skylight reflection effects in the above-water measurement of diffuse marine reflectance: comment. *Appl. Opt.*, **39**, No. 9, 1379–1381.

Kusmierczyk-Michulec, J., Schulz, M., Ruellan, S., Krueger, O., Plate, E., Marks, R., Leeuw, G. de, Cachier, H. (2001) Aerosol composition and related optical properties in the marine boundary layer over the Baltic Sea. *J. Aerosol Science*, **32**, 933–955.

Lafon, V., Froidefond, J. M., Castaing, P. (2002) SPOT shallow water bathymetry of a moderately turbid tidal inlet based on field measurements. *Rem. Sens. Env.*, **81**, 136–148.

Lahet, F., Forget, P., Ouillon, S. (2001a) Application of a colour classification method to quantify the constituents of coastal waters from *in situ* reflectances sampled at satellite sensor wavebands. *Int. J. Rem. Sens.*, **22**, No. 5, 909–914.

Lahet, F., Ouillon, S., Forget, P. (2001b) Colour classification of coastal waters of the Ebro river plume from spectral reflectances. *Int. J. Rem. Sens.*, **22**, No. 9, 1639–1664.

Leathers, R. A., Roesler, C. S., McCormick, N. J. (1999) Ocean inherent optical property determination from in-water light field measurements. *Appl. Opt.*, **38**, No. 24, 5096–5103.

Lee, Z., Carder, K. (2000) Band-ratio or spectral-curvature algorithms for satellite remote sensing? *Appl. Opt.*, **39**, No. 24, 4377–4380.

Lee, Z., Carder, K. L. (2002) Effect of spectral band numbers on the retrieval of water column and bottom properties from ocean color data. *Appl. Opt.*, **41**, No. 12, 2191–2201.

Lee, Z., Carder, K. L., Hawes, S. K., Steward, R. G., Peacock, T. G., Davis, C. O. (1994) Model for the interpretation of hyperspectral remote sensing reflectance. *Appl. Opt.*, **33**, 5721–5732.

Lee, Z., Carder, K. L., Mobley, C. D., Steward, R. G., Patch, J. S. (1998) Hyperspectral remote sensing for shallow waters: 1. A semi-analytical model. *Appl. Opt.*, **37**, No. 27, 6329–6338.

Lee, Z., Carder, K. L., Mobley, C. D., Steward, R. G., Patch, J. S. (1999) Hyperspectral remote sensing for shallow waters: 2. Deriving bottom depths and water properties by optimisation. *Appl. Opt.*, **38**, No. 18, 3831–3843.

Lee, Z., Carder, K. L., Chen, R. F., Peacock, T. G. (2001) Properties of the water column and bottom derived from Airborne Visible Infrared Imaging Spectrometer (AVIRIS) data. *J. Geophys. Res.*, **106**, No. C6, 11639–11651.

Leppaenen, J.-M., Pitkaenen, H., Savchuk, O. (1997) Eutrophication and its effect on the Gulf of Finland. In: *Proceedings of the Final Seminar of the Gulf of Finland Year 1996, March 17–18, 1997, Helsinki* (Juha Sarkkula, Ed.), pp. 31–49.

Levenberg, K. (1944) A method for the solution of certain non-linear problems in least squares. *Quant. Appl. Math.*, **2**, 164–168.

Lin, C. S. (2001) Characteristics of laser-induced inelastic-scattering signals from coastal waters. *Rem. Sens. Env.*, **77**, 104–111.

Lin, S., Borstad, G. A., Gower, J. F. R. (1984) Remote sensing of chlorophyll in the red spectral region. In: *Elsevier Oceanography Series*, Vol. 38, *Remote Sensing of Shelf Sea Hydrodynamics* (C. J. C. Nihoul, Ed.). Amsterdam: Elsevier, pp. 317–336.

Loisel, H., Morel, A. (1998) Light scattering and chlorophyll concentration in Case I waters: a reexamination. *Limnol. Oceanogr.*, **43**, 847–858.

Loisel, H., Morel, A. (2001) Non-isotropy of the upward radiance field in typical coastal (Case 2) waters. *Int. J. Rem. Sens.*, **22**, No. 2 & 3, 275–295.

Loisel, H., Stramski, D. (2000) Estimation of the inherent optical properties of natural waters from the irradiance attenuation coefficient and reflectance in the presence of Raman scattering. *Appl. Opt.*, **39**, No. 18, 3001–3011.

Loisel, H., Stramski, D., Mitchell, B. G., Fell, F., Fournier-Sicre, V., Lemasle, B., Babin, M. (2001) Comparison of the ocean inherent optical properties obtained from measurements and inverse modelling. *Appl. Opt.*, **40**, No. 15, 2384–2397.

Longuet-Higgins, M. S. (1962) The statistical geometry of a random surface. *Proc. of Symp. Appl. Math.*, Hamburg: Max-Planck Soc., Vol. 1, pp. 104–143.

Machu, E., Ferret, B., Garcon, V. (1999) Phytoplankton pigment distribution from SeaWiFS data in the subtropical convergence zone south of Africa: a wavelet analysis. *Geophys. Res. Letters*, **26**, No. 10, 1469–1472.

Maritorena, S., Morel, A., Gentili, B. (1994) Diffuse reflectance of oceanic shallow waters: influence of water depth and bottom albedo. *Limnol. Oceanogr.*, **39**, No. 7, 1689–1703.

Maritorena, S., Morel, A., Gentili, B. (2000) Determination of the fluorescence quantum yield by oceanic phytoplankton in their natural habitat. *Appl. Opt.*, **39**, No. 36, 6725–6737.

Maritorena, S., Siegel, D. A., Peterson, A., R. (2002) Optimization of a semi-analytical ocean color model of global-scale applications. *Appl. Opt.*, **41**, No. 15, 2705–2714.

Marks, R., Kruczalak, K., Jankowska, K., Michalska, M. (2001) Bacteria and fungi in air over the Gulf of Gdansk and Baltic Sea. *J. Aerosol Sci.*, **32**, 237–250.

Marquardt, D. W. (1963) An algorithm for least-squares estimation of non-linear parameters. *J. Int. Soc. Appl. Math.*, **11**, No. 2, 36–48.

Marshall, B. R., Smith, R. C. (1990) Raman scattering and in-water ocean optical properties. *Appl. Opt.*, **29**, 71–84.

Martsynkevich, L. (1970) Distribution of slopes of elementary facets of a wind-roughened water surface. *J. Meteor. Hydrol.*, **10**, 41–55.

Matthews, A. M., Duncan, A. G., Davison, R. G. (2001a) Error assessment of validation techniques for estimating suspended particulate matter concentration from airborne multispectral imagery. *Int. J. Rem. Sens.*, **22**, No. 2 & 3, 449–469.

Matthews, A. M., Duncan, A. G., Davison, R. G. (2001b) An assessment of validation techniques for estimating chlorophyll-*a* concentration from airborne multispectral imagery. *Int. J. Rem. Sens*, **22**, No. 2 & 3, 429–447.

Maul, G. A. (1985) *Introduction to Satellite Oceanography*. Dordrecht, The Netherlands: Martinus Nijhoff, pp. 423.

McClain, C. R., Yen Eueng-nan (1994) *CSZC Bio-optical Algorithm Comparison. Case Studies for the SeaWiFS Calibration and Validation*. P. I. NASA Tech. Memo. 104566. Vol. 13. Greenbelt, MD: NASA, 52 pp.

Mobley, C. D., Sundman, L. K., Boss, E. (2001) Phase function effects on oceanic light fields, *Appl. Opt.*, **41**, No. 6, 1035–1057.

Moore, G. F., Aitken, J., Lavender, S. J. (1999) The atmospheric correction of water colour and the quantitative retrieval of suspended particulate matter in Case II waters: application to MERIS. *Int. J. Rem. Sens.*, **19**, No. 9, 1713–1733.

Mordasova, N. V., Venzel, M. V. (1994) Specific features of spatial distribution of the phytoplankton pigments and biomass in the White Sea in summer. In: *Comprehensive Studies of the White Sea Ecosystem. Collected papers* (V. V. Sapozhnikov, Ed.). Moscow: VNIRO, pp. 83–92 (in Russian).

Morel, A. (1980) Optical properties of pure water and sea water. In: *Optical Aspects of Oceanography* (N. G. Jerlov & Nielsen Steeman, Eds). London: Academic Press, pp. 1–24.

Morel, A. (1988) Optical modelling of the upper ocean in relation to its biogeneous matter content (Case I waters). *J. Geophys. Res.*, **93**, 10749–10768.

Morel, A. (1990) Optics of marine particles and marine optics. In: *Particle Analysis in Oceanography* (S. Demers, Ed.). Berlin: Springer-Verlag, pp. 145–175.

Morel, A., Ahn, Yu-Hwan (1991) Optics of heterotrophic nanoflagellates and ciliates: a tentative assessment of their scattering role in oceanic waters compared to those of bacterial and algal cells. *J. Mar. Res.*, **49**, 177–202.

Morel, A., Gentili, B. (1991) Diffuse reflectance of oceanic waters: its dependence on sun angle as influenced by the molecular scattering contribution. *Appl. Opt.*, **30**, No. 30, 4427–4438.

Morel, A., Gentili, B. (1996) Diffuse reflectance of oceanic waters. III. Implication of bidirectionality for the remote sensing problem. *Appl. Opt.*, **35**, No. 24, 4850–4862.

Morel, A., Maritorena, S. (2001) Bio-optical properties of oceanic waters: a reappraisal. *J. Geophys. Res.*, **106**, No. C4, 7163–7180.

Morel, A., Prieur, L. (1977) Analysis of variations in ocean colour. *Limnol. Oceanogr.*, **22**, 709–722.

Morel, A., Voss, K. J., Gentili, B. (1995) Bidirectional reflectance of oceanic waters: a comparison of modeled and measured upward radiance fields. *J. Geophys. Res.*, **100**, 13143–13150.

Moullamaa, Yu. A. (1964) *Atlas of Optical Characteristics of Wind-roughened Marine Surface.* Tartu: Valgus Publ. Co., 531 pp.

Multi-author (1983) *Optics of the Ocean.* Vol. 1. Moscow: Nauka, 360 pp.

Multi-author (1987) *Present State of the Ladoga Lake Ecosystem* (N. P. Petrova, Ed.). Leningrad: Nauka Publishing Co., 210 pp.

Multi-author (1996) Ecological status of water bodies and water ways within the Neva River Basin (A. F. Alimov & A. K. Frolova, Eds). St. Petersburg: Publication of the Russian Academy of Sciences, 225 pp.

Multi-author (2000) *Remote Sensing of Ocean Colour in Coastal, and other Optically-Complex Waters. Reports of the International Ocean-Colour Coordinating Group* (S. Sathyendranath, Ed.). Dartmouth, Nova Scotia: IOCCG, 140 pp.

NASA (1993) *Technical Memorandum No. 104566,* Vol. 3 (S. B. Hooker & E. R. Firestone, Eds). Washington, DC: NASA, 21 pp.

Nelder, J. A., Mead, R. (1965) A simplex method for function minimization. *Comput. J.*, **7**, 308–313.

Neville, R. A., Gower, J. F. R. (1977) Passive remote sensing of phytoplankton via chlorophyll *a* fluorescence. *J. Geophys. Res.*, **83**, 3487–3493.

Neyelov, I. A., Oumnov, A. A. (1997) A model of the Neva River Bay ecological system. In: *The Neva River Bay: Experience in Modelling* (V. V. Menshoutkin, Ed.). St. Petersburg: Nauka, pp. 215–268 (in Russian).

Ohde, T., Siegel, H. (2001) Correction of bottom influence in ocean colour satellite images of shallow water areas of the Baltic Sea. *Int. J. Remote Sens.*, **22**, No. 2 & 3, 297–313.

O'Reilly, J. E., Maritorena S., Mitchell, B. G., Siegel, D. A., Carder, K. L., Garver, S. A., Kahru, M., McClain, C. (1998) Ocean color chlorophyll algorithms for SeaWiFS. *J. Geophys. Res.*, **103**, No. C11, 24937–24953.

Palmer, K. F., Williams, D. J. (1974) Optical properties of water in the near infrared. *J. Opt. Soc. Am.*, **64**, 1107–1110.

Parslow, J. S., Hoepffner, N., Doerffer, R., Campbell, J. W., Schlittenhardt, P., Sathyendranath, S. (2002) Case 2 Ocean-colour applications. In: *Remote Sensing of Ocean Colour in Coastal and Other Optically-Complex Waters* (S. Sathyendranath, Ed.). IOCCG Report No. 3, pp. 93–114.

Payne, R. E. (1972) Albedo of the sea surface. *J. Atmos. Sci.*, **29**, 959–970.

Pernetta, J. C., Milliman, J. D. (1995) *Land–Ocean Interactions in the Coastal Zone Implementation Plan*. IGBP Report No. 33, 215 pp.

Petrova, N. A. (1990) *Phytoplankton Successions with the Anthropogenic Eutrophication of Large Lakes*. Leningrad: Nauka, 200 pp. (in Russian).

Petzold, T. J. (1972) *Volume Scattering Functions for Selected Ocean Waters*. Scripps Institute of Oceanography Ref. 72-28, University of California, San Diego, 79 pp.

Pinkerton, M. H., Trees, C. C., Aiken, J., Bale, A. J., Moore, G. F., Barlow, R. G., Cummings, D. G. (1999) Retrieval of near surface bio-optical properties of the Arabian Sea from remotely sensed ocean colour data. *Deep-sea Res., Part II*, **46**, 549–569.

Pope, R. M., Fry, E. S. (1997) Absorption spectrum (380–700 nm) of pure water. II. Integrating cavity measurements. *Appl. Opt.*, **36**, No. 33, 8710–8723.

Porto, S. (1966) Angular dependence and depolarization ratio of the Raman effect. *J. Opt. Soc. Am.*, **56**, 1585–1589.

Pozdnyakov, D. V., Kondratyev, K. Ya., Bukata, R. P., Jerome, J. H. (1998) Numerical modeling of natural water colour: implications for remote sensing and limnological studies. *Int. J. Rem. Sens.*, **19**, No. 10, 1913–1932.

Pozdnyakov, D. V., Lyaskovsky, A. V. (2001) Remote sensing of natural waters: a numerical study of water leaving radiance as influenced by diurnal variations in solar illumination and near-surface wind force. *Earth Obs. Rem Sens.*, **1**, 1–8 (in Russian).

Pozdnyakov, D. V., Lyaskovsky, A. V. (1998) A comparative assessment of water quality retrieval algorithms for SeaWiFS data processing. *Proceedings of the Fifth International Conference on Remote Sensing for Marine and Coastal Environments, San Diego, California, 5–7 October 1998*, Vol. 1. Ann Arbor, MI: Environmental Research Institute of Michigan, pp. 142–148.

Pozdnyakov, D. V., Lyaskovsky, A. V. (1999) A comparison analysis of water quality retrieval algorithms for Case II waters. *Earth Obs. Rem Sens.*, **1**, 70–78 (in Russian).

Pozdnyakov, D. V., Lyaskovsky, A. V., Tanis, F. J., Lyzenga, D. R. (1999) Modelling of apparent hydro-optical properties and retrievals of water quality in the Great Lakes for SeaWiFS: a comparison with in situ measurements. *Proceedings of the IGARSS 99 Conference, Hamburg, Germany, 27 June–1 July, 1999*, Vol. II, pp. 1143–1147.

Pozdnyakov, D., Bakan, S., Grassl, H. (2000a) *Atmospheric Correction of Colour Images of Case I Waters – a Review*. Report No. 308. Max-Planck Institute for Meteorology, 33 pp.

Pozdnyakov, D., Bakan, S., Grassl, H. (2000b) *Atmospheric Correction of Colour Images of Case II Waters – A Review*. Report No. 308. Max-Planck Institute for Meteorology, 53 pp.

Pozdnyakov, D. V., Lyaskovsky, A. V., Grassl, H., Pettersson, L. (2001) Assessment of bottom albedo impact on the accuracy of retrieval of water quality parameters in the coastal zone. *Earth Obs. Rem Sens.*, **6**, 3–10 (in Russian).

Pozdnyakov, D. V., Lyaskovsky, A. V., Grassl, H., Pettersson, L. (2002a) Numerical modelling of transspectral processes in natural waters: implications for remote sensing. *Int. J. Rem. Sens.*, **23**, 1581–1607.

Pozdnyakov, D. V., Lyaskovsky, Grassl, H., Pettersson, L. (2002b) Investigation by means of numerical modeling of water colour formation in the coastal zone. *Earth Obs. Rem Sens.*, **2**, 24–37 (in Russian).

Press, W. H., Teukolsky, S. A., Vettering, W. T., Flannery, B. P. (1992) *Numerical Recipes in C: The Art of Scientific Computing*, 2nd edn. New York: Cambridge University Press, 452 pp.

Prieur, L., Sathyendranath, S. (1981) An optical classification of coastal and oceanic waters based on the specific spectral absorption curves of phytoplankton pigments, dissolved organic matter, and other particulate materials. *Limnol. Oceanogr.*, **26**, 671–689.

Reynolds, C. S. (1984) The ecology of fresh water phytoplankton. In: *Cambridge Studies in Ecology*. London: Cambridge University Press, pp. 1–150.

Reynolds, R. A., Stramski, D., Mitchell, B. G. (2001) A chlorophyll-dependent semi-analytical reflectance model derived from field measurements of absorption and backscattering coefficients within the Southern Ocean. *J. Geophys. Res.*, **106**, No. C4, 7125–7138.

Roesler, C. S., Perry M. (1995) *In situ* phytoplankton absorption, fluorescence emission, and particulate backscattering spectra determined from reflectance. *J. Geophys. Res.*, **100**, C7, 13279–13294.

Roesler, C. S., Perry, M. J., Carder, K. (1989) Modelling *in situ* phytoplankton absorption from total absorption spectra in productive inland marine water. *Limnol. Oceanogr.*, **34**, 1510–1523.

Romankevich, E. A. (1977) *Geochemistry of Organic Matter in the Sea*. Moscow: Nauka, 908 pp. (in Russian).

Roumyantrev, V. A., Drabkova, V. G. (Eds) (1999) *The Gulf of Finland under Conditions of Anthropogenic Impact*. St. Petersburg: RAS & North-Baltic Marine Fund, 363 pp.

Ruddick, K. G., Ovidio, F., Rijkeboer, M. (2000) Atmospheric correction of SeaWiFS imagery for turbid coastal and inland waters. *Appl. Opt.*, **39**, No. 6, 897–912.

Ruddick, K. G., Gons, H. J., Rijkeboer, M., Tilstone, G. (2001) Optical remote sensing of chlorophyll *a* in Case 2 waters by use of an adaptive two-band algorithm with optimal error properties. *Appl. Opt.*, **40**, No. 21, 3575–3585.

Sathyendranath, S. (2000) General introduction. In: *Remote Sensing of Ocean Colour in Coastal, and Other Optically-complex Waters* (S. Sathyendranath, Ed.). IOCCG Report No. 3, pp. 5–21.

Sathyendranath, S., Platt T. (1997) Analytic model of ocean color. *Appl. Opt.*, **36**, No. 12, 2620–2629.

Sathyendranath, S., Platt, T. (1998) Ocean-colour model incorporating transspectral processes. *Appl. Opt.*, **37**, 2216–2227.

Sathyendranath, S., Prieur, L., Morel, A. (1989) A three-component model of ocean colour and its application to remote sensing of phytoplankton pigments in coastal waters. *Int. J. Rem. Sens.*, **10**, No. 8, 1373–1394.

Sathyendranath, S., Stuart, V., Irwin, B. D., Maass, H., Savidge, G., Gilpin, L., Platt, T. (1999) Seasonal variations in bio-optical properties of phytoplankton in the Arabian Sea. *Deep Sea Res., Part II*, **46**, 633–653.

Sathyendranath, S., Cota, G., Stuart, V., Maass, H., Platt, T. (2001) Remote sensing of phytoplankton pigments: a comparison of empirical and theoretical approaches. *Int. J. Rem. Sens.*, **22**, No. 2 & 3, 249–273.

Savchuk, O., Wulf, F. (1996) Biogeochemical transformations of nitrogen and phosphorus in the marine environment. In: *Coupling Hydrodynamic and Biogeochemical Processes in*

Models for the Baltic Proper. System Ecology Contributions, No. 2, Stockholm University, 79 pp. (in Russian).

Schiller, H., Doerffer, R. (1999) Neural network for emulation of an inverse model–operational derivation of Case II water properties from MERIS data. *Int. J. Rem. Sens.*, **20**, No. 93, 1735–1746.

Schroeder, Th., Schaale, M., Fell, F., Fischer, J. (2002) Atmospheric correction algorithm for satellite data over Case-I waters. In: *Proceedings of the Seventh International Conference on Remote Sensing for Marine and Coastal Environments, Miami, Florida, 20–22 May, 2002*, pp. 123–127.

Schwander, H., Kaifel, A., Ruggaber, A., Koepke, P. (2001) Spectral radiative-transfer modelling with minimized computational time by use of a neural-network technique. *Appl. Opt.*, **40**, No. 3, 331–335.

Shifrin, K. S. (1983) *Introduction to the Optics of the Ocean*. Leningrad: Gidrometeoizdat, 278 pp. (in Russian).

Shifrin, K. S. (1988) *Physical Optics of Ocean Water*. New York: American Institute of Physics, 230 pp.

Shifrin, K. S., Zolotov, I. G. (2000) On the consideration of the slope distribution of the sea surface elements when analyzing remote sensing data. *J. Geoph. Res.*, **104**, 8025–8033.

Siegel, H. (1984) Some remarks on the ratio between the upward irradiance and nadir radiance just beneath the sea surface. *Beitraege zur Meereskunde*, **51**, 75.

Siegel, H., Michaels, A. (1996) Quantification of non-algal attenuation in the Sargasso Sea: implications for biogeochemistry and remote sensing. *Deep Sea Res., Part II*, **43**, 321–345.

Siegel, H., Gerth, M., Beckert, M. (1997) Variation of specific optical properties and their influence on measured and modelled spectral reflectances in the Baltic Sea. *Ocean Optics XII, Proc. SPIE Int. Soc. Opt. Eng.*, **2963**, 526–531.

Siegel, H., Gerth, M., Neumann, T., Doerffer, R. (1999) Case studies of phytoplankton blooms in coastal and open waters of the Baltic Sea using Coastal Zone Color Scanner data. *Int. J. Rem. Sens.*, **20**, No. 7, 1249–1264.

Siegel, D. A., Wang, M., Maritorena, S., Robinson, W. (2000) Atmospheric correction of satellite ocean color imagery: the black pixel assumption. *Appl. Opt.*, **39**, No. 21, 3582–3591.

Skouratov, S. (1997) Influence of the Pinatubo eruption on the aerosol optical depth in the Arctic in the summer of 1993. *Atmospheric Res.*, **44**, 125–132.

Smirnov, A., Villevalde, Y., O'Neil, N. T., Royeer, A., Tarussov, A. (1995) Aerosol optical depth over the oceans: analysis in terms of synoptic air mass types. *J. Geophys. Rev.*, **100**, No. D8, 16639–16650.

Smith, R. C., Baker, K. S. (1981) Optical properties of the clearest natural waters (200–800 nm). *Appl. Opt.*, **20**, 177–184.

Smyth, T. J., Groom, S. B., Cummings, D. G., Llewellyn, C. A. (2002) Comparison of SeaWiFS bio-optical chlorophyll-a algorithms with the OMEX II programme. *Int. J. Rem. Sens.*, **23**, No. 11, 2321–2326.

SNNS, Stuttgart Neural Network Simulator (1995) *User Manual, Version 3.1*. University of Stuttgart, Institute for Parallel and Distributed Performance Systems. Anonymous ftp: Informatik.uni-stuttgart.de (129.69.211.2).

Sogandares, F. M., Fry, E. S. (1997) Absorption spectrum (340–640 nm) of pure water. I. Photothermal measurements. *Appl. Opt.*, **36**, 8699–8709.

Sokolik, I. N., Toon, O. B. (1999) Incorporation of mineralogical composition into models of the radiative properties of mineral aerosol from UV to IR wavelengths. *J. Geophys. Res.*, **104**, No. D8, 9423–9444.

Solyankin, E. V., Zozoulya, S. A., Krovnin, A. C., Maslennikov, V. V. (1994) Theromohaline structure and hydrodynamics in the White Sea in the summer of 1991. In: *Comprehensive Studies of the White Sea Ecosystem. Collected Papers* (V. V. Sapozhnikov, Ed.). Moscow: VNIRO, pp. 8–25 (in Russian).

Sosik, H. M., Mitchell, B. G. (1995) Light absorption by phytoplankton, photosynthetic pigments and detritus in the California Current System. *Deep Sea Res.*, **42**, No. 10, 1717–1728.

Stramski, D. (1994) Gas microbubbles: an assessment of their significance to light scattering in quiescent seas. *Ocean Optics XII, Proc. SPIE Int. Soc. Opt. Eng.*, **2258**, 704–710.

Stramski, D., Kiefer, D. A. (1991) Light scattering by microorganisms in the open ocean. *Prog. Oceanog.*, **28**, 343–383.

Stramski, D., Reynolds, R. A., Karhu, M., Mitchell, B., Greg, A. (1999) Estimation of particulate organic carbon in the ocean from satellite remote sensing. *Science*, **285**, 239–242.

Stramski, D., Bricaud, A., Morel, A. (2001) Modeling the inherent optical properties of the ocean based on the detailed composition of the planktonic community. *Appl. Opt.*, **40**, No. 18, 2929–2945.

Stramski, D., Sciandra, A., Clausure, H. (2002) Effects of temperature, nitrogen and light limitation on the optical properties of the marine diatom *Thalassiosira pseudonana*. *Limnol. Oceanogr.*, **47**, No. 2, 392–403.

Sturm, B. (1993) CZCS processing algorithms. In: *Ocean Colour: Theory and Applications in a Decade of CZCS Experience*. Netherlands: ESA, pp. 95–116.

Sturm, B., Zibordi, G. (2002) SeaWiFS atmospheric correction by an approximate model and vacarious calibration. *Int. J. Rem. Sens.*, **23**, No. 3, 489–501.

Sugihara, S., Kishino, M., Okami, N. (1984) Contribution of Raman scattering to upward irradiance in the sea. *J. Oceanogr. Soc. Japan*, **40**, 397–404.

Tam, A. C., Patel, C. K. N. (1979) Optical absorption of light and heavy water by laser opto-acoustic spectroscopy. *Appl. Opt.*, **18**, 3348–3358.

Tanis, F. J., Marshall, E. (1989) Spatial and spectral characterization of acid rain stress in Canadian lakes, *Proc. IGARSS 89 Conf., Vancouver, Canada*, pp. 234–238.

Tassan, S. (1994) Local algorithms using SeaWiFS data for the retrieval of phytoplankton pigments, suspended sediment, and yellow substance in coastal waters. *Appl. Opt.*, **33**, No. 12, 2369–2378.

Tett, P., Kennaway, G. M., Boon, D., Mills, D. K., O'Connor, G. T., Walne, A. W., Wilton, R. (2001) Optical monitoring of phytoplankton blooms in Loch Striven, a eutrophic fjord. *Int. J. Rem. Sens.*, **22**, No. 2 & 3, 339–358.

Thiemann, S., Kaufmann, H. (2002) Lake water quality monitoring using hyperspectral airborne data – a semi-empirical multisensor and multitemporal approach for the Mecklenburg Lake District, Germany. *Rem. Sens. Env.*, **81**, 225–237.

Tikhonov, A. N., Arsonin, V. Ya. (1979) *Methods of Solution to Incorrect Problems*. Moscow: Nauka Press, 180 pp. (in Russian).

Tolk, B. L., Han, L., Rundquist, D. C. (2000) The impact of bottom brightness on spectral reflectance of suspended sediments. *Int. J. Rem. Sens.*, **21**, No. 11, 2259–2268.

Ulloa, O., Sathyendranath, S., Platt, T. (1994) Effect of the particle size distribution on the backscattering ratio in seawater. *Appl. Opt.*, **33**, No. 30, 7070–7077.

Vasilkov, A. P., Burenkov, V. I., Ruddick, K. G. (1999) The spectral reflectance and transparency of river plume waters. *Int. J. Rem. Sens.*, **20**, No. 13, 2497–2508.

Viljanen, M., Rumyantsev, V., Slepukhina, T., Simol, H. (1996) Ecological state of Lake Ladoga. In: *Karelia and St. Petersburg*. Jyvaskyla: Joensuu University Press, pp. 107–128.

Vodacek, A., Green, S. A., Blough, N. V. (1994) An experimental model of the solar-stimulated fluorescence of chromophoric dissolved organic matter. *Limnol. Oceanogr.*, **39**, 1–11.

Vohen, H. (1997) Laboratory measurement of angular distribution of light scattered by phytoplankton and silt. *Limnol. Oceanogr.*, **42**, 307–318.

Wang, M. (1999a) Atmospheric correction of ocean color sensors: computing atmospheric diffuse transmittance. *Appl. Opt.*, **38**, No. 3, 451–455.

Wang, M. (1999b) Validation study of the SeaWiFS oxygen A-band absorption correction: comparing the retrieved cloud optical thickness from SeaWiFS measurements. *Appl. Opt.*, **38**, No. 6, 937–944.

Wang, M. (1999c) A sensitivity of the SeaWiFS atmospheric correction algorithm: effects of spectral bands variations. *Remote Sens. Environ.*, **67**, 348–359.

Wang, M. (2000) *The SeaWiFS Atmospheric Correction Updates* Vol. 9, NASA Technical Memo. 2000-206892, SeaWiFS Postlaunch Technical Report Series (S. B. Hooker & E. R. Firestone, Eds). Greenbelt, MD: NASA Goddard Space Center, pp. 57–63.

Wang, M. (2002) The Rayleigh lookup tables for the SeaWiFS data processing: accounting for the effects of ocean surface roughness. *Int. J. Rem. Sens.*, **23**, No. 13, 2693–2702.

Wang, M., Franz, B. A., Barnes, R. A., McClain, C. R. (2001) Effects of spectral bandpass on SeaWiFS-retrieved near-surface optical properties of the ocean. *Appl. Opt.*, **40**, No. 3, 343–348.

Warrender, C. E., Augusteijn, M. F. (1999) Fusion of image classification using Bayesian techniques with Markov random fields. *Int. J. Rem. Sens.*, **20**, No. 10, 1987–2002.

Waters, K. J. (1995) Effects of Raman scattering on water-leaving radiance. *J. Geophys. Res.*, **100**, No. C7, 13151–13161.

Whitlock, C. H., Poole, L. R., Ursy, J. W. (1981) Comparison of reflectance with backscatter for turbid waters. *Appl. Opt.*, **20**, 517–522.

Whitte, W. G., Whitlock, C. H., Harriss, R. C. (1982) Influence of dissolved organic materials on turbid water optical properties and remote sensing reflectance. *J. Geophys. Res.*, **87**, 441–446.

Wozniak, S. B. (1999) Modelling of the experimental factors influence on solar irradiance reflectance and transmittance through the wind-roughened sea surface. *Proc. SPIE Int. Soc. Opt. Eng.*, **2963**, 84–89.

Wu, J. (1988) Bubbles in the near-surface ocean: a general description. *J. Geophys. Res.*, **93**, 587–590.

Yacobi, Y. Z., Gitelson, A., Mayo, M. (1995) Remote sensing of chlorophyll in Lake Kinneret using high-spectra-resolution radiometer and Landsat TM: spectra features of reflectance and algorithm development. *J. Plankton Res.*, **17**, No. 11, 2155–2173.

Yan, B., Chen, B., Stamnes, K. (2002) Role of oceanic air bubbles in atmospheric correction of ocean color imagery. *Appl. Opt.*, **41**, No. 12, 2202–2212.

Zepp, R. G., Schlotzhauer, P. F. (1981) Comparison of photochemical behavior of various humic substances in water. III. Spectroscopic properties of humic substances. *Chemosphere*, **10**, 479–486.

Zhang, X., Lewis, M., Johnson, B. 1998. Influence of bubbles on scattering of light in the ocean. *Appl. Opt.*, **37**, No. 27, 6525–6536.

Zibordi, G., Berton, J.-F. (2001) Relationships between Q-factor and seawater optical properties in a coastal region. *Limnol. Oceanogr.*, **46**, No. 5, 1130–1140.

Index

Printing: Mercedes-Druck, Berlin
Binding: Stein+Lehmann, Berlin